The Sex Imperative

An Evolutionary Tale of Sexual Survival

The Sex Imperative

An Evolutionary Tale of Sexual Survival

Kenneth Maxwell

Plenum Press • New York and London

Library of Congress Cataloging in Publication Data

Maxwell, Kenneth E., date.
The sex imperative: an evolutionary tale of sexual survival / Kenneth Maxwell.
        p.   cm.
    Includes bibliographical references (p.   ) and index.
    ISBN 0-306-44649-9
    1. Sex (Biology)—Evolution. I. Title.
QP251.M417   1994   94-96
575'.9—dc20   CIP

QP
251
.M417
1994

ISBN 0-306-44649-9

© 1994 Kenneth Maxwell
Plenum Press is a Division of Plenum Publishing Corporation
233 Spring Street, New York, N.Y. 10013-1578

Printed in the United States of America

Acknowledgments

This book was made possible by scientists, too numerous to name, whose research findings in biology and related sciences built a vast body of knowledge that provided the story of sex and its relation to inheritance, evolution, and the human condition. I hope that my account of their work does them justice.

I want to acknowledge the contributions of my agent, Michael Snell, and especially Plenum editor Linda Greenspan Regan, who, through the give and take over details and content, served to greatly sharpen the focus of the book, for which I am deeply grateful.

Contents

I. Origins

II. Connections

III. Sex Organs

IV. Beyond Nature

Part I

Origins

Chapter 1

The Biological Bang

The creation of the universe began with the Big Bang. No one knows where the energy came from, and no one knows what touched it off. A few billion years later there was another bang, just as spectacular, just as mysterious, and just as creative—the Biological Bang.

The universe was created about 15 billion years ago, give or take a few billion years. The age of our galaxy, the Milky Way, is thought to be close to 12 billion years, and that of the sun and formative solar system about 4.6 billion years. The age of meteorites and rocks in the earth's crust can be estimated by determining the relative abundance of radioactive precursors of the lead, uranium, and thorium in present-day rocks. By calculating their rate of radioactive decay, it is estimated that the planet earth was formed about 4.5 billion years ago. Less than a billion years later, something mysterious happened on earth that was destined to bring about profound changes on the planet: the appearance of life.

The earliest clearly identified fossils of complex organisms are found in sediments in western Australia, which were determined to be nearly 3.5 billion years old. William Schopf at the Center for the Study of Evolution and the Origin of Life at the University of California, Los Angeles, discovered a great diversity of cyanobacteriumlike microorganisms in a region of northwestern Western Australia underlain by a sequence of sedimentary and volcanic rock 30 kilometers thick and about 3.0 to 3.5 billion years old. Sheath-enclosed colonial unicells were found in a layer of sedimentary rocks formed about 3.465 billion years ago. Included were eight previously undescribed species of filamentous microorganisms. The find gives strong evidence of the presence of oxygen-producing photosynthesizers early in Earth history. The conclusion is inescapable that microbial life had evolutionary roots predating, perhaps substantially, 3.465 billion years ago. Simple living organisms had to originate earlier than that, probably around 4 billion years ago.

No one knows for sure how it happened. Prophets of different cultures and at various times handed down their inspired views of creation. The Greek philosophers took a shot at it. Aristotle and his pupil, Theophrastus, and later Lucretius, speculated in a vague way on the origin of living things. Aristotle believed that living things could and did originate spontaneously from inorganic matter without the need for preexisting life. Theophrastus was a doubter, and Lu-

The cloud-whirled planet Earth, as seen in this view of the Pacific Ocean from Apollo 11. A traveler arriving from space would see that the earth's surface is predominantly water, an ideal solvent and support for the creation of life. Courtesy of the National Aeronautics and Space Administration.

cretius concluded that "the earth has ceased to create life," but the doctrine of spontaneous generation persisted as the prevailing view for 2000 years until Louis Pasteur laid the idea to rest in 1860 by a simple but ingenious laboratory experiment with crooked-neck flasks containing culture media.

One of the first in modern times to offer a plausible theory of the origin of life was Charles Darwin. He wrote to a close friend, botanist Joseph Hooker, in a letter dated February 1, 1871,

> . . . But if (and oh what a big if) we could conceive in some little warm pond, with all sorts of ammonia and phosphoric salts, light, heat, electricity, etc., present, that a protein compound was chemically formed, ready to undergo still more complex changes, at the present day such matter would be instantly devoured or absorbed, which would not have been the case before living creatures were formed.

Darwin's "little warm pond" came surprisingly close to the primordial soup many of today's scientists envision as the conditions under which life originated. But there is ample room for speculation. A troublesome feature of tracing our ancestral origins according to the warm pond theory is the fact that the fossil record does not reveal a simple beginning. Bacteria and algae, the most primitive organisms for which there is evidence in the fossil record, are not simple aggregations of atoms and molecules. Single-cell organisms are microscopic factories that use complex biochemistry, much of which is even now incompletely understood or not fathomed at all.

An event that brought the mystery of life's origin closer to understanding was staged in the laboratory of Nobel laureate Harold Urey at the University of Chicago. Among other accomplishments, Urey worked out theories on the formation of the planets, and suspected that life is common throughout the universe. Alexander Oparin, a Soviet biochemist, had proposed in an article published in 1924 and expounded further in his book, *Origin of Life* (1936), a theory that life could have arisen spontaneously on earth by purely random processes of chemistry and physics. A similar view arrived at independently by the British scientist J. B. S. Haldane was published in 1929. The idea appealed to Urey, who included it in his book, *The Planets* (1952). Urey's graduate student, Stanley Miller, set out to create some of the rudiments of life by setting up a miniature environment of the kind that was presumed to have prevailed on the primitive earth. He put some water in a flask, making a miniature sea. There was a reason for using a tiny ocean as a starting point. Water is the liquid of life, the essence of the living process.

Protoplasm consists mostly of water; it forms most of our flesh and blood. The dissolved salts in living tissue are similar to those in seawater, not surprising if one assumes that life was born in the sea. Humans and other terrestrial vertebrates reveal sea-dwelling ancestry in the early fetal stage by the presence of gill slits and other embryonic similarities to primitive aquatic animals more or less representative of the ancestral type. We begin life in a cradle of fluid. Shortly after the fertilized human ovum becomes implanted in the wall of the uterus, membranes appear that are not part of the embryo but form a sort of aquarium for the developing fetus so that it can grow as its fishlike ancestors did before they abandoned their aquatic life for the adven-

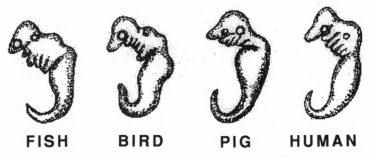

FISH　　BIRD　　PIG　　HUMAN

Early-stage embryos of fish, bird, pig, and human are very much alike, evidence, according to one theory, of a common ancestral origin in the sea. Drawing by Vicki Frazior.

ture on land. Even the embryo itself is mostly fluid, for water forms most of our body throughout life.

Miller added an atmosphere of hydrogen, ammonia, and methane, and heated the flask to vaporize some of the water. Energy is needed to bring about organic change, so Miller circulated the witch's brew—lethal to any present form of life—in the gaseous state through a tube past an electrical discharge and artificial sunlight. After a week, Miller found that the brew contained several amino acids—the building blocks of proteins—several organic acids, aldehydes, and hydrogen cyanide. These are all potential precursors of more complicated chemicals in living organisms. Miller published the results of his experiment in 1953. He had taken a giant step in inner space.

Other scientists produced still more life-related molecules under various conditions that simulated those presumed to have existed on the primordial earth: adenosine triphosphate (ATP), an important energy-transfering molecule present in, and vital to, all living cells; nucleic acids, the primary building blocks of the basic genetic materials deoxyribonucleic acid (DNA) and ribonucleic acid (RNA); and porphyrin complexes, pigments related to chlorophyll, which is the key to photosynthesis and essential to all plant and animal life.

But it is a long journey from salt to cell. A mixture of molecules in a test tube is not a living organism. A bacterium, an amoeba, or an alga is enormously complex. To be sure, polyamino acids, when in contact with water, tend to form structures that have many of the characteristics of living cells. Sidney Fox and his co-workers at Florida State University in 1958 put some amino acids on a piece of lava rock, baked them in an oven, then washed the rock in salt water. The amino acids had taken on the shape of tiny spheres that moved and were able to proliferate by a kind of budding. Fox called them "proteinoid (proteinlike) microspheres." They were not living organisms, but displayed superficially some characteristics of life.

Because no one knows how life originated, scientists proposed various theories, almost all of them hotly debated. A widely accepted theory is an elaboration of the warm pond idea. The curtain opens on a

PLANKTON AND OTHER SURFACE WATER ALGAE

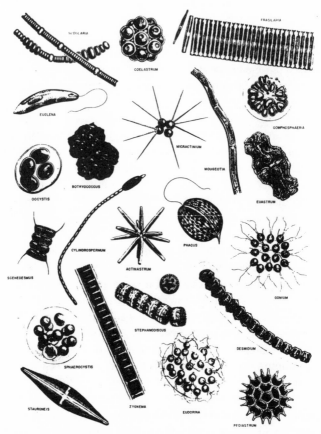

Microorganisms commonly found in fresh surface water. These algae are infinitely more complex than the first living organisms. Courtesy of C. M. Palmer, from *Algae in Water Supplies*, U.S. Public Health Service, 1959.

scenario that began 4 billion years ago. Volcanic eruptions and other sources within the earth enriched the earth's atmosphere, giving the air a gaseous composition different from the relative abundance of cosmic elements in the universe. If, as Oparin, Haldane, and Urey thought, the atmosphere of the earth was predominantly hydrogen, a heritage from the nearest star—the sun—reactions would have occurred that would not be possible in today's high-oxygen atmosphere. It is now thought, however, that most of the hydrogen, the lightest element, would have been blown away by the solar wind. As the earth gradually cooled, some of the water in the gaseous envelope of the earth condensed into puddles and steaming ponds. A dense fog contained warm vapors of water, methane, ammonia, and small amounts of deadly hydrocyanic acid. These were potential precursors of life, but more was needed.

Energy is needed to bring about change, and there was an ample supply on the primitive earth. The atmosphere, presumed to be low in oxygen, did little to impede energetic radiations emanating from nuclear reactions in the sun. The cauldron of life on earth was blasted with ultraviolet radiation, gamma rays, and other highly energetic radiations from the sun and space with a force that would wipe out present forms of life. There was the impact of cometary meteors, radioactivity from the crust of the earth, blazing heat from the sun, and radiant warmth from the oven of mother earth. And the slowly cooling, but still warm, surface of the wet earth was convulsed with bolts of lightning and sonic shocks of thunder of shattering intensity for eons on end.

The environment was not everywhere the same. Shallow seas held various concentrations of dissolved minerals. Salt ponds had various compositions, and some were warmer than others. Some of the larger bodies of water had tidal action, but it was probably in the warmer shallow fringes that the miracle of life took place. The forces of attraction that cause atoms to stick together with various degrees of tenacity gave rise to several kinds of organic compounds, including an occasional nitrogen base. From the nitrogen bases

Life may have originated in warm, shallow, tidal seas similar to this, except under more severe conditions of the early earth, bombarded with high-intensity radiation, jolting electrical discharges, and swirling hot gases.

nucleic acids were born—the building blocks for the macromolecules of genetic material. Some of the macromolecules stuck together in jellylike microglobules—still too small to cast a shadow—picking up smaller molecules and inorganic salts along the way. After a long time, perhaps several million years, thin films appeared around some of these ionic and molecular aggregates, forming celllike globules of jelly.

Then something happened that was as remarkable and portentous as the creation of the universe. Some of the macromolecules in the jelly globules had the ability to catalyze chemical reactions on their own, making the jelly blobs miniature chemical factories, now called protoplasm, from the Greek *protos*, meaning *first*, and *plasma*, meaning *form*. Nucleic acids were born that were capable of reproducing themselves in a suitable environment. One of the results was that some of the globules of protoplasm split up, forming baby globules exactly like their parents. It may have happened many times in many places, but if so, nearly all of the globules of protoplasm must have been destroyed in the ever-changing, hostile environment of dried up ponds, shallow seas, torrential flooding, rampant volcanism, deadly volcanic ash, scorching heat, unrelenting bombardment by comets and asteroids, and possibly predation by competing strains of protoplasm that had already learned to engulf others for nourishment. One of the globules thrived, giving rise to a dynasty of globules—the ancestors of all living things.

Where and how all this happened is a matter of speculation. Sidney Fox and Klaus Dose in 1972 proposed the so-called *thermal theory* whereby heat was a significant source of free energy on the primitive earth and life must have originated under relatively dry conditions instead of in the open sea, as many people supposed. Fox became a strong proponent of the idea that proteins were the first form of life, preceding genetic material, as opposed to the more popular view that genetic material came first. Harold Morowitz, a professor of biology and natural philosophy at George Mason University, is another proponent of the theory that protoplasm preceded replicating genetic material. Fox's ideas developed into the *bubbles* theory in which protenoids in the form of bubbles floated on the sea surface and became the progenitors of life. A theory more plausible to most scientists was the proposal of Christian de Duve, co-winner of the Nobel Prize in Physiology and Medicine in 1974. His theory, published in 1991, favored an environment for the beginning of life consisting of strongly acidic sulphurous hot springs—deep-sea hot water geysers called *hydrothermal vents* where there were simple organic compounds and phosphate that would give rise to *protometabolism*. A related theory is that life originated in a volcanic lake. A persuasive argument for the subterranean hot spring theory is that the formative organisms were protected in the ocean depths from the murderous surface bombardments that are presumed to have lasted for a billion years or more, so violent that no living thing would be able to survive.

A controversial theory was the proposal that life or its progenitors came from outer space. A Scottish physicist, William Thompson, later Lord Kelvin, was one of the first to suggest that parts of living organisms, including seeds, might be scattered through space, a possibility he mentioned during a speech in 1881. He had suggested in 1871 that life may have been brought to earth in meteorites. A more fully developed theory was promoted by a Swedish chemist, Svante Arrhenius, who in an article written in 1903, and in a book, *Worlds in the Making* (1908), maintained that life exists throughout the universe and came to earth in the form of living spores. He called his idea *panspermia*, meaning *seeds everywhere*. Later advocates of life from outer space included Sir Fred Hoyle, a professor of astronomy at Cambridge University, and Chandra Wickramasinghe, a professor of mathematics and astronomy at University College in Wales. They wrote several books in which they expounded views that were harshly criticized by other scientists. Nobel Laureate Francis Crick and Leslie Orgel, a scientist doing work on prebiotic chemistry at the Salk Institute, La Jolla, California, proposed a theory in 1973 that they called *directed panspermia* by which living forms may have arrived on earth on spacecraft from another planet. The theory was explored further by Crick who offered it as a topic for discussion in a book, *Life Itself* (1981), but the idea received little support. Among others who speculated on the idea, the British jack-of-all-sciences, J. B. S. Haldane had mentioned the possibility in 1954. However, none of the theories about life on earth coming from outer space are widely accepted in scientific circles. While conceding that if life could originate on earth, it could just as easily originate on other worlds, most scientists say that claims of an extraterrestrial origin merely sidestep the fundamental question of how it happened. Whether life came from outer space or from a pond, the process culminating in the first spark of life must have been much the same.

A theory proposed by A. G. Cairns-Smith, a chemist at the University of Glasgow, Scotland, is that the earliest organisms were clay minerals. According to this theory, genetic information was in the irregularities in the crystal structure of the clay. The minerals were able to catalyze chemical reactions, thus controlling the organic chemistry of the organisms. Replication occurred simply by crystal fission. A related theory is that clay served as templates for the formation of living organisms. The idea that clay had a role in the origin of life dates back to J. Desmond Bernal, who published his lectures of the late 1940s in a book, *The Physical Basis of Life*. The clay theory is reminiscent of the biblical and other early stories in which the first human was made from clay.

A widely held view is that a form of genetic material called RNA was the start of it all. Genetic engineering revealed that RNA can act as a true enzyme—a biochemical catalyst. A simplified scenario is an "RNA world" in which life began with the chance appearance of a catalytic, self-replicating RNA that evolved further into complex organisms. A contrary view held by Christian de Duve, a Belgian cellular biologist, Robert Shapiro, a chemist at New York University, and others, is that the chance synthesis of even a single molecule of RNA would be so improbable that it would "verge on the miraculous." In an effort to avoid this scientific quagmire, others proposed that RNA arose spontaneously and became established by *genetic takeover*, replacing a more primitive enzyme, or it developed in a system containing a group of simple, mutually catalyzing molecules. In any case, there came into being an ancestral cell, the common ancestor of all living things.

The key to continuity of life on earth is nature's scheme for genetic communication. The way living organisms transmit genetic information from parents to offspring is so simple in outline, and so universal throughout the spectrum of life, that the method was almost certainly adopted at a very early stage in the creation of life. All organisms carry the genetic information inside their cells in the form of giant molecules called *nucleic acids*. There are two closely related nucleic acids referred to by the acronyms DNA and RNA, for deoxyribonucleic acid and ribonucleic acid. Molecules of DNA contain regularly repeated structures suggestive of vertebrae in an enormously long, continuously curving spine in the shape of a double helix, likened to a spiral stairway. Attached to each side of this curving backbonelike structure are side groups called *nitrogen bases*, of which there are four kinds identified by their initials A, G, C, and T. These stand for adenine, guanine, cytosine, and thymine. Another base, U, for uracil, replaces thymine in RNA. The bases are the alphabet of a genetic language used by all living things from microbes to people. The sequence of the letters (the bases) forms a code that is understood by every cell in the body, and as genetic engineering proved, can be understood even by cells of totally unrelated organisms.

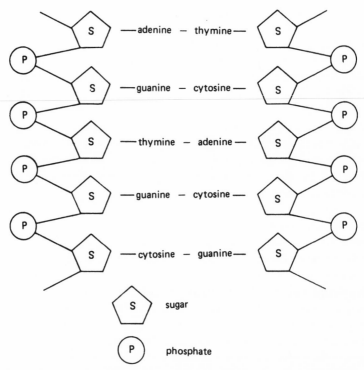

The double helix of DNA. Each half consists of a sugar–phosphate (S-P) backbone holding nitrogen bases, adenine (A), thymine (T), guanine (G), and cytosine (C).

Although all organisms functioned in accordance with the genetic vocabulary since the origin of life, it took human ingenuity to decode it. After revelation of the structure of DNA by James Watson and Francis Crick in 1953, biochemists made a furious effort to find out how it functioned. Many contributed, and the answer came surprisingly quickly. By 1961, Marshall Nirenberg at the National Institutes of Health and Gobind Khorana at the University of Wisconsin, working independently, showed that the nitrogen bases coded in groups of three bases, referred to as *triplets* or *codons*. Each combination of triplets signals that a specific amino acid is to be incorporated into a protein. For example, the triplet UUU is a codon for phenylalanine, and GUA is a codon for valine. The entire code for all the amino acids was quickly worked out.

The richness of an organism's genetic vocabulary depends on its physiological and behavioral needs, and this is generally related to its size. Simple organisms do not need as much information in their genetic vocabulary as do more complex organisms. A virus may have about 5 to 10 thousand bases in its genetic instructions, whereas a much more complicated bacterium may have as many as 4 million base pairs. Humans may contain 50 million to 100 million base pairs, but one estimate is that only 2 to 5% of them are functional, the rest being arranged in "nonsense" codons.

When sex began is not certain, but the appearance of sexuality in the most primitive organisms suggests that it began early in the evolution of life. At the very beginning, there was no need for it. The master macromolecules of genetic material supervised the growth and functions of the other cellular molecules. They reproduced by budding or splitting in two. The offspring received master molecules exactly like the ones their parent cells had possessed, and these molecules could usually be counted on to supervise the physiology and behavior of each generation of offspring in the old familiar way. But over time, conditions on earth changed. There were drastic changes in climate, radiation from the sun, temperature, and salt concentrations because of drying and flooding. The cells came under increasing stress to find ways to nourish themselves and survive the cruelty of the elements. Many of them perished. Living things were in danger of becoming extinct before making a mark on the world.

Fortunately, as it turned out, the master molecules of genetic material, now known as DNA and RNA, were not always perfect. Occasionally an atom or a molecule zigged when it should have zagged, and when that happened, it caused the cell to function differently. This kind of miscue is commonly called a *mutation*. Most of the mutations were lethal to the organism or its offspring. A very small fraction of a percentage of the mutations were beneficial, causing our ancestral cells to change in ways that made them more adaptable to threatening conditions or in other ways improved their chance of survival. Over millions of years, many kinds of cells appeared in various forms and shapes and in different environments. Some of them stuck together in the form of colonies, leading to division of labor and specialization. Some cells were good at gathering nourishment, some were good at reproduction, some of them were good at rowing into uncharted seas with microscopic oars called cilia, whereas others could dive and swim by using whiplike extrusions called flagella. The chemistry of the cells became equally diverse. Eventually millions of different kinds of multicellular organisms appeared, the products of a process that has come to be called *evolution*.

But long before multicellular organisms made their appearance, a strange thing happened that accelerated evolution and improved the survival odds of a particular dynasty, or species, and contributed to the relatively rapid appearance of a multiplicity of living forms. Two free-living cells came together and passed parts of their substance from one to the other, including some of their genetic material. Sex was invented.

Chapter 2

Why Sex?

The question "Why sex?" has long been a discussion topic, from the whimsical treatment by James Thurber and E. B. White in *Is Sex Necessary?* to a serious piece by John Maynard Smith, *What Use is Sex?* in the *Journal of Theoretical Biology*. The question of whether sex is good, bad, or indifferent may never be resolved to the satisfaction of everyone, but the majority view is that there is an evolutionary advantage to sex.

The first living things to emerge from the tortured planet during the creation of life were almost certainly sexless. But sex must have appeared on the scene soon afterward, for exchange of genetic material in one form or another exists in the simplest organisms. Even viruses, those strange parasitic bundles of molecules that hover in the twilight zone between the living and the nonliving, have their special brand of sex. Thus sex was an early innovation that was retained through the evolutionary process from throbbing aggregations of molecules in the primordial slime to the complicated consortium of human cells engaged in what is called the joy of sex.

Sex means different things to different people. Depending on the context, it means male or female, sexuality, sensual pleasure, copulation, reproduction, or to a biologist, the transfer of genetic material from one organism (usually a male) to another organism (usually a female) in a way that makes the progeny genetically different from the parents. Sex, to the biologist, can be simply recombinations of genetic material within cells. Sex in that sense is not the same as reproduction.

There can be sex without reproduction, as we shall see, and reproduction without sex.

Sex was invented by microbes, and for whatever reason, the habit caught on. Throughout the evolutionary history of life from microbes to primates, organisms devised many unique, often bizarre, ways of doing it. Some species can reproduce with or without sex, and a relative few eschew the practice altogether, managing to reproduce for generation after generation without sex. But most organisms, including all the higher animals, are sexual creatures, unable to reproduce at all without sex. Is there an advantage other than ephemeral pleasure? Why sex at all? A clue to the answer is found in how evolution works.

The idea that species undergo change from one to another was kicked around for centuries without having much influence on the way people viewed nature. Most people thought that plants and animals were created by divine design to be the way they were, predictable and immutable. The first person to formulate a full-blown theory of evolution with an explanation of how it could happen, and to forthrightly promote the idea, was Jean Baptiste de Lamarck, a former French Legionnaire with a flair for natural history.

Jean Baptiste Pierre Antoine de Monet, chevalier de Lamarck, as he called himself, was born in 1744 as the 11th child of an impoverished family of aristocrats whose traditional professions had been in the army and the church. Jean was picked for the church against his will, so when his father died, he promptly went off to enlist in the infantry, where he served with distinction marked with exceptional bravery and was rewarded with a commission. Following an injury, he figuratively traded his musket for a microscope and pursued the study of medicine and then switched to natural history, eventually becoming botanist to the king, then professor of invertebrate zoology at the Museum of Natural History in Paris. Lamarck made important and lasting contributions in both botany and zoology. He founded and organized modern invertebrate zoology, and was the first to use the terms *vertebrate* (an animal with a backbone) and *invertebrate* (an animal without a backbone). But he is most remembered for a flaw in his theory of evolution, described in one his books, *Zoological Philosophy*, published in 1809, the year Charles Darwin was born.

Lamarck contended that species are not constant, as most people thought, but change over time, and that they do so by parents passing to their offspring characteristics that are acquired through some need. For example, a short-necked prehistoric ancestor of the giraffe would have to stretch its neck higher and higher to reach increasingly scarce foliage. Animals with stretched necks would transmit this physical attribute to their progeny, and so on for succeeding generations, the process continuing for as long as giraffes needed longer necks for their nutritional well-being.

Lamarck was not the first to come up with the theory that evolution was brought about by so-called *inheritance of acquired characteristics*. A decade earlier, Charles Darwin's grandfather, Erasmus Darwin, a prominent physician, was a firm believer in evolution and anticipated Lamarck in espousing the idea that plants and animals evolved by reason of environmental influences. But Erasmus liked to express his philosophy in didactic verse that was not easy to interpret. He was regarded as a dilettante and did not cut much of a figure in scientific circles. Lamarck was the first of the serious naturalists to gain wide recognition as an evolutionist. Unfortunately, he is not remembered as much for his considerable accomplishments and for championing evolution against formidable and sometimes vitriolic opposition as for the fact that his theory of evolution by acquired characteristics did not hold up in the light of Charles Darwin's exhaustive studies and the inescapable conclusion that evolution was driven by natural selection of random variations.

Charles Robert Darwin, like his father, the son of a prominent physician, was a likeable sort. He enrolled in medical school at Edinburgh but became a dropout and took up studying for the clergy at Cambridge University to assuage his father's anger at the prospect of his becoming nothing more than a "sporting man." Fortunately, one of his professors at Cambridge detected the rudiments of genius in the young man and arranged for his appointment as a naturalist on the now famous expedition of the H.M.S. *Beagle*, which embarked from Liverpool on December 27, 1831. Darwin's observations were published in detail in *A Naturalist's Voyage of the Beagle* (1839) and *On the Origin of Species by Means of Natural Selection, or the Preservation of Favoured Races in the Struggle for Life* (1859), usually known simply as *The Origin of Species*. Having seen wide variability in physical and behavioral characteristics within species, he became convinced that evolution depends on natural selection of members of the population that have traits most favorable for survival and that the traits are inherited. For instance, during the *Beagle*'s visit to the Galapagos Islands, Darwin noted that there were at least 21 species of birds and that they existed nowhere else. Several species of finches occupied the islands, each one with physical characteristics that differed from all the rest in a way that enabled it to occupy its own special niche. One species had a long, narrow beak suitable for reaching into cracks and crevices of the bark of trees and shrubs for insects. Another species of finch had a short, stout beak suitable for cracking hard seeds. Darwin believed that these specialized forms evolved from a common ancestor species that, like all plants and animals, had individuals that differed slightly from one another. The long-beaked ones were especially adept at finding food where no other bird could reach it and were able to survive when others might go hungry. The heavy-beaked finches could thrive when less endowed birds would fail to feed their young.

For evolution to work, the favorable characteristics would have to be inherited, and Darwin knew from personal experience with domestic animals that many characteristics could be passed on to their progeny by selective breeding. In Darwin's view, natural selection operated in nature just as effectively. He wrote,

> . . . can we doubt (remembering that many more individuals are born than can possibly survive) that any individuals having any advantage, however slight, over others, would have the best chance of surviving and procreating their kind? This preservation of favorable variations and the rejection of injurious variations, I call Natural Selection.

Whereas Lamarck a generation earlier had theorized that evolution was caused by the inheritance of characteristics that were acquired by the parents through accommodation to some need such as stretching their necks for food, Darwin contended that there are naturally occurring random variations among organisms. When these are inherited, the progeny that receive beneficial characteristics will be the ones most apt to survive, and they will pass the beneficial qualities on to their progeny.

Another naturalist, Alfred Russel Wallace, who had studied wildlife in several parts of the world, hit on the idea of evolution by natural selection before Darwin announced his theory. Wallace wrote a description of natural selection while confined to his hammock with an attack of malaria in Borneo, and by an irony of fate, sent the manuscript to Darwin for review. Darwin's friends suggested that the two naturalists make a joint announcement, so papers on the theory submitted by each were published in the *Journal of the Linnaean Society* in 1858.

Natural selection is a two-edged sword. It also weeds out individuals that for various reasons are poorly equipped to survive. Darwin devoted a chapter in his *Origin of Species* to the Struggle for Existence in which he referred to Herbert Spencer's expression "survival of the fittest," allowing that it might be more accurate than his own description. But a piece of the puzzle was missing. If chance variations made the difference between species, what caused variations in the first place? And how were they inherited? Stung by criticism, Darwin began to have doubts that his theory was the pat answer to evolution that he had hoped for when he wrote his book on the origin of species. Nagging questions threw an ominous shadow over the validity of his theory. Nothing can be more stimulating and at the same time more frustrating to a scientist than to have an incomplete answer. Charles Darwin was faced with a dilemma. His theory of variation and natural selection did not explain the source of the variations. Without the missing piece of the puzzle, he was left, in effect, with only half a theory. As time went on, and detractors of his theory grew strident, Darwin became acutely aware of the need to find the missing piece of the puzzle. He struggled with the problem for years and then came up with a remarkably good guess but not quite the right answer. Ironically, part of his answer came uncomfortably close to the explanation offered more than a half century earlier by the French evolutionist, Lamarck, whose ideas he had previously dismissed as "rubbish."

Darwin's new theory, which he called *pangenesis*, was described in *Variation in Animals and Plants under Domestication*, published in 1868. The theory of pangenesis drew partly from ancient concepts of heredity dating back to the time of the early Greeks and ideas held by others with various modifications. Darwin assumed that in every cell of the body there are invisible hereditary particles, which he called *gemmules*. He said that gemmules migrate to *germ cells* from all parts of the body, making it possible for the progeny to inherit the characteristics of the parent. There are present, Darwin postulated, a large number of different kinds of gemmules, and because every cell in every part of the body throws off its own peculiar kind of gemmules, the mixing of gemmules explains much that can be observed in inheritance. Some of the gemmules can remain dormant, he said, and their activation accounts for the reversion to ancestral characteristics. For instance, a brown cow may have a latent characteristic for black color that may show up in its calf. Moreover, gemmules could be modified by external conditions, resulting in newly acquired traits that could be transmitted to offspring. In the final part of his pangenesis theory, Darwin said that the tissues of the body, being directly affected by new conditions, are able to "throw off modified gemmules, which are transmitted with their newly acquired peculiarities to the offspring." In other words, new characteristics could be acquired by direct effect of the environment. Darwin's shot in the dark was a near miss of the present concept of genes and mutations. We know now that mutations have environmental causes, such as exposure to x rays or chemicals, as well as internal causes, such as virus infections or glitches of unknown origin in the biochemical machinery of the cells during formation of germ cells. Darwin died without knowing the complete answer, although an important clue discovered by a methodical monk, Gregor Mendel, that might have stimulated Darwin's agile mind lay collecting dust on the shelves of about 120 European libraries for the last 16 years of his life.

Gregor Johann Mendel was born to misfortune and failure. He grew up in hardship and poverty as the son of an Austrian peasant. He struggled desperately for an education, and finally had to settle for an Augustine monastery in Brünn (Brno in the modern Czech Republic), where he became a monk but failed as a priest and had a nervous breakdown. Sent away to the University in Vienna, he failed his exams—twice—and ended up in 1854 as a substitute teacher without a credential. Finally, his brilliant discovery—the greatest of the century—was ignored as inconsequential or just plain wrong by the science cognoscenti of the day and went unnoticed for 40 years. Mendel had discovered that inheritance followed a mathematical pattern. Some of the "charakteres" disappeared in hybrid progeny and reappeared in later generations. For example, he crossed two strains of pea plants, one with round seeds and one with wrinkled seeds. When the hybrid produced seeds, they were all round. Mendel called the factor for round seeds *dominant*, and the factor for wrinkled seeds *recessive*. But the round seeds of the hybrid were not all genetically the same as those of the parent, because when the hybrid's round seeds were planted, three fourths of the resulting plants produced round seeds, and one fourth of them produced wrinkled seeds. A Danish botanist, Wilhelm Johannsen, proposed the word *gene*, from a Greek word meaning *race* or *kind* for the factors responsible for the different traits. It has been claimed that Mendel fudged his data and was lucky in choosing pea plants for his studies because they represent one of the simplest examples of inheritance. No matter, because his conclusions have withstood the test of time and tens of thousands of confirming experiments.

We pick up the thread of the story of inheritance a generation later when Hugo de Vries, a Dutch botanist, rediscovered Mendel's laws of inheritance and made a serendipitous discovery strictly his own. While out for a walk in the countryside near Amsterdam, he noticed something unusual about two plants in a patch of evening primrose. The plants were so different from the others that de Vries took them to be new species. Follow-up experiments established the fact that new characteristics could suddenly appear in a population of plants or animals. De Vries called them *mutations*. Most of the transformations he saw were not actually new species nor even what we now call gene mutations, but he demonstrated once and for all that there could be rapid changes in inheritable traits such as the color and form of flowers or the shape, size, and arrangement of leaves.

Meanwhile, Walther Flemming, a German anatomist, by using dyes, revealed some of the inner workings of animal cells during cell division. Material scattered through the cell nucleus strongly absorbed a dye. He called the material chromatin, from the Greek *chroma*, for color. During cell division, the chromatin coalesced into threadlike objects that came to be called *chromosomes* (colored bodies). Flemming described his work in a book, *Cell Substance, Nucleus, and Cell Division*, published in 1882. A German biologist, August Weismann, suggested that the chromosomes contained a hereditary element he called *germ plasm* that would divide during cell fission, maintaining the hereditary mechanism intact. He further proposed that half of the germ plasm went to the egg and half to the sperm and that the original quantity was restored during fertilization. His theory proved to be basically correct, but it left no room for variations and therefore did not explain the missing part of the story of evolution.

Variation could be explained logically only by de Vries's mutations, which were brilliantly confirmed and some of their causes established by Hermann Muller who studied under the geneticist Thomas Hunt Morgan at Columbia University. Muller found in 1919 that mutations in the fruit fly, *Drosophila*, could be increased by heat. Then in 1926, while at the University of Texas, he found that x rays, being more energetic than heat, greatly increased the mutation rate. He had demonstrated that mutations, and therefore variations in traits, could be created by human intervention. For his work on x rays, Muller was awarded the 1946 Nobel Prize in medicine and physiology.

Soon after Muller demonstrated the mutagenic effects of x rays, Albert Blakeslee, at the Carnegie Station for Experimental Evolution, began experimenting with chemicals. He found that colchicine, an alkaloid obtained from the autumn crocus, could double the chromosome number in cells of plants, resulting in a condition called polyploidy, also changing the plant's characteristics. Although not a true gene mutation, the finding opened the floodgates to trials with other chemicals, many of which proved to be true mutagens (physical or chemical agents that cause mutations). The first chemical found to have this action on animals was nitrogen mustard, evaluated by J. M. Robson and Charlotte Auerbach of the University of Edinburgh at the beginning of World War II and tested in the United States for treatment of cancer. At about the same time in Scotland, mustard gas and nitrogen mustard were found to cause mutations in *Drosophila*, and in Germany, urethane was found to cause chromosome breakage and chromosome rearrangement in plants.

It had by now become apparent that mutations, which could have a variety of causes, were the original source of the elusive variations desperately pursued by Charles Darwin. And it was beginning to become clear that although some mutations could improve the chance of survival, variations caused by other mutations could be so extreme that the mutated individuals would not survive. For instance, mutations that cause cancer, extreme physical deformities, physiological malfunctions, or any of a large number of genetic diseases are often lethal or disabling.

But biologists have long recognized another source of variation as well—transfer of genetic material from one individual to another by the act of sex. Darwin was intimately familiar with the methods of artificial selection used to improve livestock and horticultural crops and saw that they had much in common with natural selection. He had made extensive study of animal and plant breeding to improve the stocks, practices that were common among wealthy landholders and stock raisers who found an increasingly lucrative market for farm products in the burgeoning industrial society. Darwin and other members of his family were fanciers of pigeons, birds noted for having a fascinatingly wide variety of inheritable traits. He and others saw how the qualities of offspring could be manipulated by selective breeding to produce desired characteristics in the progeny. It was easy to see that the same selective breeding techniques could be used to produce progeny with undesirable characteristics if anyone were foolish enough to want them.

Mutations are sudden changes in heritable genetic material. They become evident when new physical or physiological characteristics appear in an individual that were not present in previous generations. A typical mutation is one that arises from an alteration in a single gene, causing it to code for a different protein or proteins in a way that may have profound effects, good or bad, on the makeup of the organism. A single gene mutation is called a *point mutation*. Mutations can also arise from gross alterations in the structure, location, presence, or absence of chromosomes. These kinds of alterations are called *chromosomal mutations*.

Mutations are the initial source of genetic *variation*. Some shuffling of genes is brought about when gametes are formed and chromosomes exchange pieces by crossing over, and by so-called jumping genes in defiance of Mendel's law. Sex is the major player in survival of the fittest because the transfer of genes by sex recombines genetic material in millions of possible combinations that result in many different phenotypes. Mutations are relatively infrequent; sex and recombination of genes is a daily occurrence. The need for constant stirring of the genetic pot by sexual recombination of genes is the *sex imperative*. For instance, the early global environment was almost certainly one of violence and constant change that placed organisms in great peril. If all the organisms were genetically identical clones, new conditions such as sudden changes in temperature, moisture, or food supply might wipe out the entire population. By sexually producing a motley batch of offspring possessing a wide spectrum of survival capability from fits to misfits, many lines would certainly die out through mortality or failure to reproduce. But some offspring would be better equipped to survive changes in the environment. Most mutations probably have little effect one way or the other, a large number of mutations are lethal or incapacitating, and probably a small fraction are beneficial.

August Weismann, a German biologist described by Harvard's Ernst Mayr as ". . . one of the great biologists of all time," was one of the first to point out the importance of sex to the process of evolution. Weismann, born in 1834, studied medicine and served as a surgeon in the Austrian army and then became professor of zoology at the University of Freiburg-im-Breisgau. The development of eye trouble made it impossible for him to use a microscope, so he turned to theory, especially regarding various aspects of evolution and inheritance. He wrote in 1886, referring to fertilization, ". . . two groups of heredity tendencies are combined. . . . The object of this is to create those individual differences out of which natural selection produces new species." He continued, "I do not know of any meaning that can be attributed to sexual reproduction other than the creation of heritable individual characteristics upon which selection may work." In other words, the sex imperative is the process of providing the variety of genetic combinations that give natural selection a chance to operate. Sir Francis Galton, a first cousin of Charles Darwin, recognized the importance of sex as a cause of variation. He thought that desirable qualities of humans could be increased by proper breeding, and used the name *eugenics* for the study of that method of human improvement. Charles Darwin did not recognize the full impact of sex on variation because he believed that the characteristics of parents were blended in the progeny. He did not have the advantage of knowing about Mendel's results demonstrating segregation and recombination of genes in progeny and their influence on dominant and recessive characteristics.

Weismann's view of the crucial role of sex as a source of the unlimited variations in populations became widely accepted. Most biologists see sex as a powerful driving force in evolution. Hermann Muller, a pioneer in work on mutations, wrote,

> The essence of sex, then, is Mendelian recombination . . . the production, among many misfits, of some combinations that are of permanent value to the species . . . the advantage of sexual over asexual organisms in the evolutionary race is enormous.

Theodorus Dobzhansky, an eminent Russian–American geneticist, declared,

> The evolutionary advantage brought about by the appearance of sex and the emergence of new organismic entities, Mendelian populations, are enormous. . . . The appearance of sexual reproduction was perhaps the most important advance in the evolution of life.

And Ernst Mayr, a distinguished Harvard biologist, expressed a confirming corollary,

> There is a little doubt that abandoning sexuality cuts down drastically on future evolutionary options. Evolutionary lines that switch to uniparental reproduction most likely become extinct sooner or later. . . .

Some biologists take a more extreme view. Robert Burton, a British biologist, wrote, "Without the variation caused by sexual reproduction natural selection could hardly operate."

Biologists have come up with numerous theories for the function of sex, some of them described with colorful names such as The Red Queen, the tangled bank, the Vicar of Bray, Muller's ratchet, the best man, the lottery ticket, and the hitchhiker. An intriguing school of thought is that sex was a historical accident that is no longer biologically useful, but there is no way to get rid of it. One of the first to espouse a theory on the accidental origin of sex was E. C. Dougherty, who suggested in 1955 that sex originally functioned as a mechanism for DNA repair. The idea is that DNA, bombarded constantly from without and within, frequently suffers damage from mutations caused by radiation, chemicals, parasites, and slips in the cogs of the internal biochemical machinery, but recombination of genes by whatever method has a chance of repairing or mitigating the damage.

One view is that DNA repair is the only function of sex, and in most organisms the need for it no longer exists. The theory was argued by Lynn Margulis, a biologist at Boston University, and coauthor Dorion Sagan in their book, *Origin of Sex* (1986),

> We do not think there is any evidence to justify the claim that sexual organisms are more diverse and therefore better equipped to cope with the vicissitudes of existence, nor that they reproduce in a sexual fashion because this permits them to "evolve faster."

Margulis and Sagan elaborated on the idea that the origin of sex was the result of a series of historical accidents involving the repair of damaged DNA. A condensed and simplified version of their argument is a scenario in which high-intensity ultraviolet light, unimpaired by the early earth's atmosphere, imposed the life-threatening danger of DNA damage that required a repair mechanism. The cells were already equipped with enzymes for DNA splicing (removal of a section of DNA and connecting the remaining pieces) and polymerase enzymes (naturally occurring enzymes that promote DNA replication and repair). When the DNA came from an outside source, that is, another cell, sex became an important source of repair parts. When atmospheric oxygen and a build-up of ozone appeared after the proliferation of photosynthetic organisms, thus reducing the intensity of ultraviolet radiation, repair by means of DNA recombination became less urgent. But sex remained as carry-over excess baggage, no longer needed for its original purpose.

The idea that sex always serves to accelerate evolution was questioned earlier by others. George Williams, a biologist at the State University of New York, Stony Brook, contended that sex is a poor strategy where a species is so well adapted to its environment that large numbers survive. He pointed out that in animals that reproduce both sexually and asexually, sexual reproduction is an adaptation to a specific situation, such as to facilitate dispersal. This is seen in some species of aphids where generation after generation of wingless, asexual females give birth to great numbers of young by parthenogenesis (without fertilization by males) on a particular plant. When the end of the season approaches or the food supply is nearly exhausted, they suddenly start giving birth to winged male and female sexual forms that fly away to a different species of plant where they reproduce sexually. A similar alternation of sexual and asexual reproduction is seen in some parasites (for example, the protozoan that causes malaria, which achieves dispersal by spending part of its life cycle in mosquitos).

The extreme view that DNA repair is the sole function of sex is not sustainable. George Williams declared,

> . . . that synapsis was originally evolved and is still maintained as a mechanism for repairing DNA, and not as a mechanism for generating variation among descendent chromosomes, strikes me . . . as too extreme.

James Crow of the University of Wisconsin wrote,

> . . . asexual species may well have immediate evolutionary advantage, but are less successful in the long run. The prevailing view is that asexual species are less able to keep up with environmental changes than are sexual ones.

There is little doubt that sex has a repair function, but the theory that sex was an accident of nature of temporary benefit but of no further use, like an appendix, fails to explain why sex has remained the dominant feature of reproduction for several billion years despite infinite opportunities to slough it off for the more energy-efficient asexual reproduction. If during a much shorter time, snakes can lose their legs—evolutionary legacies no longer useful for their way of life—many more of the millions of organisms that cling tenaciously to sex could dump the habit in favor of a more energy-conserving way of perpetuating the species.

If sex were a useless strategy for species that regularly reproduce sexually, why did they not revert to asexual reproduction? Actually, some of them did, but relatively few. Several species of insects, a few species of fish, and even a few species of amphibians and reptiles are known to reproduce exclusively without sex. If the unlikely reversion to nonsex as an adaptation ever becomes an evolutionary trend among higher animals, it is conceivable, though so highly improbable that the chances of it occurring are almost nonexistent, that a woman somewhere could undergo a mutation enabling her to give birth parthenogenetically. *Homo sapiens* would evolve in such a milieu into a truly sexless society, regrettably reducing the status of men to eunuchoids or slaves, to the dismay of both men and women.

The most successful animals that ever existed on earth are the insects. Relatively few of the million or more species took the evolutionary path to nonsexual reproduction. Nothing stands in the way of the rest of these remarkably fecund creatures ridding themselves of a practice that is highly costly in time and energy—except for the evolutionary advantage of the sex imperative. Their survival by adaptability and evolutionary change since their appearance more than 300 million years ago is attributable in large measure to their persistent, seemingly uninhibited, sexual activity. Continuous creation of new combinations of genes and a mixture of phenotypes almost ensures the survival of some descendants even if subjected to drastically changing environments.

The dramatic effect of sexual activity on adaptability and evolution was demonstrated by microbes, as we shall see.

Chapter 3

Single Cell Sex

As I write this I am looking at a picture of a single cell—a microbe. It takes up the better part of a page, about as much as a dinosaur in the *National Geographic* on a nearby table. The big picture of a small critter is not misleading. Everyone knows that a microbe is smaller than a dinosaur. But how small? The size of single cells varies among species; the largest bacterium on record is one named *Epulopiscium fishelsoni*, found in the intestines of surgeonfish in the Red Sea. It measures 0.6 millimeter long, large enough to be visible to the unaided eye. This monster is a million times the volume of the most studied and most commonly used bacterium for experimentation, *Escherichia coli*. A typical bacterium will measure about 1 to 2 μm (microns or micrometers) long, completely invisible to the naked eye (1 μm = .000001 meter = .00004 inch). One thousand typical bacteria could be placed end to end across the period at the end of this sentence. But inside each of these invisible entities is a microcosm of enormous complexity. Part of the complexity is the mechanism for reproduction.

The most energy-efficient ways for a cell to reproduce are by *fission* or *budding*. Fission consists of simple division, for example, as seen in a rod-shaped bacterium, which will elongate to nearly twice its length then form a partition in the middle that eventually separates the cell into two daughter cells. Meanwhile, the cell's structural components double in number, giving each daughter cell a full complement of the contents of the parent cell. The generation time, or doubling time, may be as short as 10 minutes for some bacteria or for others as long as 24 hours or more. Most bacteria have a doubling time in the range of 1 to 3 hours. The common intestinal bacterium *Escherichia coli* (usually written *E. coli*) doubles about every 20 minutes. Budding, a form of reproduction commonly seen in yeast, consists of pinching off a bulblike protrusion from the parent cell. A new cell begins as a small outgrowth of the old cell; the bud enlarges, then separates. As in fission, the daughter cell gets a full complement of the parent cell's components.

In either fission or budding, the daughter cells are identical to the parent. There is conformity from generation to generation, barring an occasional mutation. These are perfect ways for organisms to perpetuate themselves in a stable environment. But when conditions change, as always happens sooner or later, organisms that have lived a comfortable life generation after generation with no genetic change may find it hard to adapt to conditions of a new environment. Many will die, perhaps all of them. Only if there are segments of the population that differ from the norm in their genetic makeup will any of them have a good chance of surviving when there is a drastic change in the environment. Genetic differences among individuals can arise by mutations, but many mutations are harmful, and only a few are beneficial. A wider spectrum of genetic types is achieved when organisms throughout the population engage in an exchange of genes—through a process called sex. When two organisms share their genes in a sexual exchange, the process is much like reshuffling a deck of cards in a way that gives each progeny a new deal, unlike the hands held by either parent. The result is a heterogeneous population in which ideally, there will always be some members more adaptable than others and more able to cope with changing conditions, giving them a greater chance of survival. Natural selection will shift the makeup of the population toward individuals that have the advantage of a more diversified genetic capability. This is the sex imperative.

When microorganisms learned to engage in a primitive kind of sex, a single cell was no longer bound to make mere carbon copies of itself. Its daughter cells could enjoy the benefits of the diversity provided by two, rather than merely one, parent. Some of the off-spring would succumb to the adversities of the environment. Starvation, excessively hot or cold temperatures, too high or too low salt concentrations, too much or too little sunlight, chemical toxins, and predators would take their toll. But other offspring would have qualities that would greatly increase the odds of coping with the environment and surviving to become parents themselves.

A primitive form of sexual union is called *conjugatory sex*, or simply *conjugation*, in which two cells come together briefly, ex-change nuclei, then go their separate ways to continue to reproduce for a few generations by fission. The single-cell ciliated protozoan *Paramecium* is a typical organism that engages in conjugation. *Paramecium* was probably not the first organism to engage in sexual repro-duction, but its behavior seems to come close to depicting a sophisticated form of original sex. This little organism, large for a single cell, has a funnel-shaped groove ending in a mouth that scoops in food as it propels itself around with its thousands of beating cilia. When it finds a mate, the two of them come together side by side, a small hole opens between them and they exchange micronuclei. There is no male–female relationship in the usual sense, although they have what is called *mating types*. If two *Paramecium* cells that are not

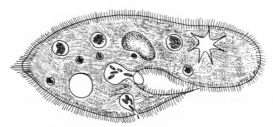

Artist's rendering of a *Paramecium*. Several species of these microscopic protozoans have active and varied sex lives. Drawing by Vicki Frazior.

complementary mating types approach each other, they soon separate. No one knows why. If complementary, they touch together at what is called the *holdfast* region near the front end of each cell. Next, near the other end, they become closely attached where a small opening develops in each cell to allow the cytoplasm of the two cells to come in contact. Each cell contains two kinds of nuclei—a large *macronucleus* that helps control the cell's physiology, and one or more small *micronuclei* that come into play in reproduction. During conjugation, a micronucleus in each cell migrates to the other cell where it fuses with a corresponding micronucleus. Conjugation is prolonged, after which the cells separate and later divide by fission.

Sex among the single cells is not always a simple matter of boy meets girl. *Paramecium* cells are meticulously choosey; their choice of mates depends on which variety within the species and which mating type within that variety the intended partner belongs to. The rule is that cells of one mating type will avoid mating with each other and will mate only with cells of another mating type in the same variety of the species they belong to. There are two sexual groups of *Paramecium* species. One group of species contains varieties that have only two mating types each. In the other group, some of the varieties have several mating types. *Paramecium aurelia* has eight varieties, each with two mating types, and another variety with only one mating type. The sex life of *Paramecium caudatum* is even more complicated with 13 varieties and 25 mating types. *Paramecium bursaria* has three varieties with four mating types each, and one variety with eight. In species having varieties with several mating types, cells will mate freely with cells of different mating types but rarely with a cell not in the same variety; however, cells of a variety that has only one mating type must mate with members of another variety, but only with certain mating types.

The common colon bacteria, *E. coli*, has three mating types that engage in a complicated transfer of genetic and other cellular material from donor cells to recipient cells. The sex life of the lowly slime mold *Physarum polycephalum*, which is found in such places as rotting logs where it feeds on other microorganisms, especially bacteria, is more sophisticated. A slime mold is an intermediate form of life that is hard to classify. In its vegetative stage it moves and behaves like an animal, but when it reproduces, it forms a spore stalk like a fungus (the fungi are a diverse group of organisms that includes molds, yeasts, and mushrooms; unlike algae, the fungi do not contain chlorophyll). *Physarum* has a hierarchy of 13 sexes, each one of which is incompatible with the one above it. If a slime mold ranks number 13, it is at the top of the heap and can mate with any other slime mold; its mitochondrial genes carried in the cytoplasm will be inherited in each mating. But if it ranks number 9, its cytoplasmic genes will be rejected by the higher ranks and will be inherited only if it mates with those ranking 9 or below. What qualifies a slime mold for a particular rank in the hierarchy is a mystery.

There are further questions about these microorganisms. If they are going to have sexes, why more than two? Laurence Hurst and his colleague William Hamilton at Oxford University puzzled over this and came up with an explanation of why some microorganisms have multiple sexes. Single cells, though mostly too small to be seen with the naked eye, are highly complex organisms, each a microscopic ecosystem. The conventional view is that the structural inclusions called organelles, the chemical communication system, the energy-using and energy-producing systems, and the reproduction machinery all perform with blissful cooperation in a benevolent cytoplasm for the common good. Not so, say Hurst and Hamilton. They picture the cellular innards as an arena for a wild and woolly free-for-all in which the cytoplasmic components—the mitochondria, plasmids, chromosomes, and other inclusions—are all looking out for themselves.

Given that the main purpose in life is to look out for number one, a battle over nutrients or a power struggle for control of cellular physiology might erupt if the organelles of two cells were intermingled when mating. The ciliates such as *Paramecium* avoid most of the potential intracellular warfare by exchanging only micronuclei containing their essential genetic information. Each partner keeps its own cytoplasm and organelles. They further insulate themselves from microwarfare by having mating types that exclude incompatible mates. They have circumvented the need for separate male and female sexes. But how about cells that mate by *fusion sex*, another way that microorganisms mate during which both nuclei and cytoplasm are exchanged?

The green alga *Chlamydomonas*, though a plant, not an animal, exemplifies organisms that engage in fusion sex. When two alga cells mate they "spill their guts," transferring parts of their cytoplasm containing organelles along with their nuclei. But unlike *Paramecium*, the green alga has two sexes, a + type (female) and a − type (male). Only the + type's organelles are inherited. Some ciliates engage in sex either way—conjugation or fusion. When they engage in conjugatory sex, in which there is an exchange only of nuclei, they have multiple mating types, and when they engage in fusion sex, in which both nuclei and cytoplasm are transferred, like the green algae, they have only two sexes. Disruptive and perhaps lethal intracellular conflict is avoided. The system that has evolved in higher animals illustrates the point that Hurst and Hamilton make about intracellular warfare. The egg or ovum is a complete cell in which its mitochondrial genes as well as nuclear genes can be inherited. A sperm, on the other hand, contains little cytoplasm and few organelles. It has little to fight for, or with, except its DNA. There is no conflict. It seems clear that the mating-type relationships of the single cells, convoluted though they may be, are a foreshadowing of the familiar relationship of higher animals. Early in the game of sex, probably 3 billion years ago, single cells invented behavior that still persists as a biological imperative in most of the higher animals including humans: avoid inbreeding; choose outbreeding (cross-breeding).

A mating type may be thought of as a large family of close relatives because paramecia of a mating type refuse to mate with other paramecia of the same mating type. Because there are several varieties of mating types, they must contend with many family groups. How these little animals find the right mate is a mystery. A theory is that each cell has a mating-type chemical on its surface or an identifying pattern of the surface structure. Another possibility is that the cilia of the same mating types adhere to each other but not to cilia of an incompatible mating type. In any case, their picky habits are a demonstration of the biological imperative that is already expressing itself: as in human societies, inbreeding is to be avoided in order to minimize the disastrous effects of progeny getting a double dose of potentially injurious genes. There is no need for a teleologic interpretation of this behavior. Darwin's natural selection adequately explains the driving force that compelled some of the most primitive organisms to establish the biological imperative to minimize inbreeding and maximize outbreeding. Manipulation of heredity by supervised breeding of animals and plants has long been both an industry and an avocation. Experiments with animals, as well as casual observations of animals and humans, verify that inbreeding is apt to cause trouble and that outbreeding can be highly beneficial. Genetics shows why.

The word *gene* comes from a Greek word meaning *descent* or *to give birth to*. The function of a gene is to provide the recipe for the amino acid sequence in a specific protein. A single bacterial cell such as the common colon bacterium, *E. coli*, contains roughly 1000 kinds of proteins totaling 2 or 3 million protein molecules. A simple bacterial cell may have as many as 3000 genes, but probably only one third of them are expressed at any particular time. In 1941, Edward Tatum and geneticist George Beadle, then at Stanford University, started work on a mold called *Neurospora crassa* in which they discovered that the function of a gene is to supervise the formation of a particular enzyme, a substance that catalyzes a biochemical reaction. This gave rise to the theory, "One gene, one enzyme" or since enzymes are proteins, "One gene, one protein." Beadle, Tatum, and another geneticist, Joshua Lederberg, shared the 1958 Nobel Prize in medicine and physiology.

Genes exist in pairs; they are paired because the chromosomes that carry them are paired. Each of the genes of a pair is called an *allele*. A pair of alleles may consist of identical genes or two genes that differ. For example, in some breeds of dogs, a pair of genes is responsible for black or red coat color. One allele of the pair (B) is responsible for black coat; the opposite allele (b), sometimes appearing as a mutant gene, is responsible for red coat. If both alleles in the dog's chromosomes are for black coat (BB), a condition called *homozygous*, the dog's coat will be black. If the alleles are homozygous for red coat (bb), the dog's coat will be red. But if the alleles for both black and red coat are present (Bb), a condition called *heterozygous*, the dog's coat will be black. Thus the gene for red coat is expressed only if both alleles for red coat (bb) are present. The black-coat gene (B) is said to be *dominant*, and the red-coat gene (b) is *recessive*.

If a mutation occurs, it can have any of several possible effects. Some mutations are neutral, having no observable anatomic or physiological effects, or none of any significance. A rare mutation might be beneficial, enabling progeny that inherit the gene to better cope with life. But mutations are more apt to be harmful. The odds of trouble from a mutated recessive gene are greatest if there is a history of mating within the circle of close relatives, many of whom may have inherited the defective gene. The offspring that receives a deleteriously mutated gene will always be in trouble if the gene is dominant. If the mutated gene is recessive, it will not be expressed in the progeny unless both parents contributed an identically mutated allele, making the offspring homozygous for that particular pair of genes. An exception is so-called *sex-linked* inheritance in which a recessive gene contributed by the mother may have no corresponding allele on the short *sex* chromosome contributed by the father. How this affects inheritance is shown by how sex (gender) is determined.

Sex is determined in humans and many other animals by how the so-called sex chromosomes designated X and Y are paired in the cells. Females have XX and males XY, but in humans, the Y chromosome is shorter than the X chromosome, so the X chromosome contains some genes without matching pairs on the Y chromosome. Those genes will be expressed in the anatomy or physiology of the organisms even if the genes are recessive. If one of the chromosome genes with no matching

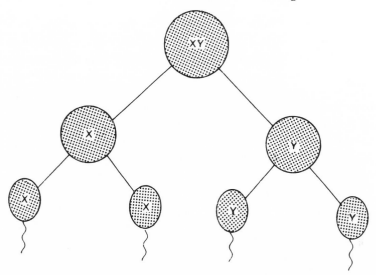

Transmission of mammalian sex chromosomes during the formation of sperm (spermato-genesis) in the testicles. Union of an egg with an X sperm results in a female (XX) embryo; union with a Y sperm results in a male (XY) embryo.

allele changes through a harmful mutation, an XY individual (male) carrying the mutated gene will be affected. A notorious case of sex-linked inheritance is that of hemophilia (bleeder's disease) in the families of European royalty. The most common form of the disease, caused by a sex-linked recessive gene, is carried on the X chromosome. Females (XX) that are heterozygous, that is, with only a single dose of the gene, do not manifest the disease because the recessive gene is countered by the dominant normal allele. But if a male child (XY) receives the gene on the X chromosome contributed by his mother, there is no corresponding allele on the short Y chromosome to counteract it, so it expresses itself fully by causing the disease. Queen Victoria was heterozygous for hemophilia, probably due to a mutation in one of her parents. She had a hemophiliac son and two healthy heterozygous daughters who married into the reigning families of Spain and Russia. Several of Queen Victoria's grandsons and great-grandsons in the royal families of those countries were hemophiliacs. She also had a healthy son, Edward VII, who carried on the British monarchy free of the hemophilia gene.

Genes mutate at widely different rates, depending on the particular gene as well as its exposure to environmental mutagens: agents such as chemicals, radiation, or heat that can cause mutations by disrupting the structure of DNA. Spontaneous mutations occur at an estimated average rate of about 10^{-5} per generation. This means that on average there is one chance in 100,000 that a given gene will mutate for each cell division. The estimate is based largely on calculations made by Thomas Hunt Morgan, who, while at Columbia University, worked out much of the genetics of the fruit fly, *Drosophila*, for which he was awarded the 1933 Nobel Prize in medicine and physiology. In human terms, it is variously estimated that each human chromosomal cell contains at least 50,000 to 100,000 genes. If we take the lower figure and assume that each cell contains 50,000 genes and that the mutation rate is 10^{-5}, it follows that one gamete (sperm or ovum) in two, or every other one, contains a newly mutated gene. Of course, not all of the mutations will be deleterious or detectable. But mutations are frequent enough to result in the common occurrence of genetic disorders. At least 5% of the children born alive have birth defects that require medical attention. About one baby in 100 is born with a defect so serious that it either dies or is seriously handicapped. But most of the defects take their toll earlier, because of failure of the eggs to implant in the uterus or of spontaneously aborted fetuses. It is estimated that one in 10 pregnancies terminates in a miscarriage, often caused by a genetic disorder of the fetus. A unicellular animal having a much smaller complement of genes, say 3000 pairs, is at great risk. A deleterious mutation would have a high probability of being lethal, especially if the genes are homozygous from inbreeding (mating of closely related individuals). Given the rapid rate of reproduction of most single-cell organisms, natural selection would give a better chance of survival to those that avoid the risks of inbreeding. Outbreeding reinforces the sex imperative by enhancing the genetic heterogeneity of the population.

In nearly all human societies, it is forbidden to mate within the immediate family. In some societies, a man is under compulsion to take a bride only within his ethnic or religious circle. Marriage of close relatives is a taboo in most human cultures, probably with only a vague understanding of the genetic reasons for it, although the more perceptive of the priests and medicine men must have noticed deleterious effects. The practice in some primitive cultures of raiding neighboring tribes for wives, usually with only minor bloodshed, may stem from an intuitive need for outbreeding. Charles Darwin, who married his first cousin, never came to grips with the problem of inbreeding (first cousins have one eighth of their chromosomes in common).

The most dramatic effects of inbreeding and outbreeding are seen in experiments on corn plants, *Zea mays*, that led to revolutionary developments in agriculture and nutrition. G. H. Shull reported in 1908 on his experiments at the Cold Spring Harbor experiment station in New York in which he self-fertilized several generations of plants, a procedure that required a special technique because corn plants normally cross-fertilize. He found that after several generations of inbreeding, the plants were progressively reduced in size and vigor (after six generations of brother–sister crosses, there may be a 50% loss of genetic variability). Meanwhile, from the inbreeding he had produced a number of pure-breeding lines, each having different characteristics. When he crossed these highly inbred lines he obtained plants that were more vigorous than their parents, even exceeding in vigor those of the original stock. People realized that *hybrid vigor*, which geneticists call *heterosis*, had promise of being put to practical use. But the procedure was too complicated and costly for corn farmers to use until methods were developed to produce hybrid seed on a commercial scale. One of the first to realize the commercial possibilities of hybrid corn seed was Henry Wallace, who studied agriculture at Iowa State University, and started his career as editor of his father's farm journal. He became secretary of agriculture under Franklin D. Roosevelt, then Roosevelt's successful running mate for vice president, and eventually became a presidential candidate. Wallace began in the 1930s marketing a double-cross type of hybrid corn seed, which, along with other commercial hybrid seeds, greatly increased the yield of the United States corn crop. He also launched a hybrid chick industry in 1942. It is said that Henry Wallace was one of the richest men ever to run for president.

Similar results of hybrid vigor are obtained with livestock, although it is more difficult, time-consuming, and costly. Besides the time it takes animals to mature, the procedure requires sperm banks and artificial insemination. Breeds of dairy cows such as Holstein, Guernsey, and Jersey are generally highly inbred strains. Experiments show that crossbreeding of dairy cows results in lower mortality of both calves and adults, fewer barren cows, and higher fertility. The first generation benefits, but the problem is that hybrids do not breed true, so the traits cannot be fixed in subsequent progeny.

Sometimes the sex imperative compels organisms to breed with such abandon that an accelerated rate of evolution by natural selection brings about swift changes in genetic makeup. This is seen in microorganisms when they suddenly evolve into virulent pathogens that are devastating to the human species. Let us do a flashback to 1885 when a German bacteriologist, Theodor Escherich, isolated a bacterium that he named *Bacterium coli* because the organism lives in the mammalian colon, including that of humans. The genus name was changed later to *Escherichia* in honor of its discoverer. The bacterium, hereafter referred to as *E. coli*, is widely used for laboratory studies and is also a commonly useful indicator of water pollution. It also has the distinction of being the first bacterium to be observed engaging in sexual reproduction. In 1946, Joshua Lederberg, a geneticist, found that by mixing strains of *E. coli*, he could get them to transfer genetic material from one strain to another, in effect crossing them. Lederberg at the time was a graduate student of Edward Tatum, a biochemist then at Yale University. Lederberg also showed in 1952 that bacteriophages—viruses that parasitize bacteria—could carry genetic material from one bacterium to another. In 1955, François Jacob and Elie Wollman at the Pasteur Institute devised a clever experiment that threw light on the way a male bacterium transfers its genetic material to the female. Their finding, discussed by Horace Freeland Judson in *The Eighth Day of Creation*, led to the so-called Spaghetti Hypothesis. While working on a strain of *E. coli* called the *Hfr* (high-frequency recombination) strain, they were able to separate the males from the females at intervals. Not all the genetic characteristics were transferred at the same time, but in piecemeal fashion. The researchers concluded that during mating the male extruded a strand of genes that had been broken during sexual contact, the location of the break depending on the length of time the mating pair was left together. Laboratory wags called the experiment "coitus interruptus."

The microbe madness seen in mating types, multiple sex, sexual

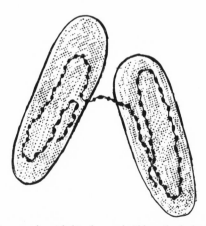

The *coitus interruptus* experiment led to the spaghetti hypothesis: bacteria do not transfer their genetic material all at once during a sexual encounter. The number of genes transferred from donor cell to recipient cell depends on the time they are in contact. Drawing by Vicki Frazior.

hierarchies, conjugation, fusion, and spaghetti sex sometimes defies rational explanation. An intuitive response is to say, "So what? Let the little critters have their microscopic sport!" Unfortunately, the sex imperative in the small world of microbes is a big cause of human misery. It plays a major role in bacteria becoming resistant to anti-microbial drugs. Many kinds of organisms have the capacity to become resistant to the toxic effects of chemicals. Populations of bacteria, fungi, insects, rats, mice, other rodents, and plants typically tend to become progressively less susceptible over time when exposed to chemicals designed to eradicate them.

Resistance to the toxic effect of chemicals is dependent on the presence of certain enzymes capable of breaking down the chemicals to less injurious substances. The ability is genetic in nature and is inherited. Some cases of resistance can be pinpointed to a single gene; in other cases several genes may come into play. A wild population of flies, for example, consists of individuals that differ in their suscep-tibility to being poisoned by DDT, a chemical that kills on contact. The flies have different physiological characteristics that are inher-ited. Most of them may die from the poison, but a few of them are apt to survive. The survivors will pass the genes for survival on to some of their offspring, whereas the least resistant individuals will not perpet-uate the genes for susceptibility because they will be dead. The process is one of Darwinian (natural) selection whereby organisms that are better endowed genetically to deal with (detoxify) toxic agents will become resistant and survive, but those that do not have the genetic endowment to cope with the toxic environment will become nonresistant and will die. Resistant individuals will pass on their genes for resistance to their progeny, rapidly building up a population of resistant organisms. Laboratory workers culturing houseflies through several generations found that their cultures became so resistant to DDT that the flies were unaffected by exposure to a 100% concentra-tion of the chemical. The sex imperative as it relates to the need for organisms to produce progeny with a variety of characteristics to ensure survival and success of the species applies to pathogenic (disease-causing) microbes as well as to other organisms.

Sexual promiscuity in microorganisms elevates their resistance to drugs used to combat infectious diseases to a new and worrisome dimension. Resistance has increased menacingly over the past half century. The reason is that microbes are remarkably generous in sharing their genes for resistance through sexual contact. According to Michael Cohen of the Centers for Disease Control in Atlanta, Georgia, "Infections that are essentially untreatable have begun to occur as epidemics both in the developing world and in institutional settings in the United States."

At first, microbe hunters were serene in their confidence that an abundant supply of new drugs would enable them to successfully battle any microbe. Until the 1980s, drug companies were churning out new antimicrobe drugs so steadily that when a pathogenic (disease-causing) organism became resistant to one, there was always another drug to take over. But "bacteria are cleverer than men," says Harold Neu, professor of medicine and pharmacology at Columbia University. The drug companies have almost stopped looking for new antimicrobials, mainly for economic reasons. It costs about $200 million to bring a new drug to market, and the greatest need is in developing countries where most of the people cannot afford drugs.

The battle between microbes and medicine is ominously one-sided because sexually active bacteria are shamelessly promiscuous. An astonishing discovery was made in 1928 by Frederick Griffith, a British scientist working in the Pathological Laboratory of the Ministry of Health. Griffith was working on *Streptococcus pneumoniae* (pneumococcus) bacteria that cause pneumonia. There are two different strains of the bacterium, one with a smooth carbohydrate coat (S), and the other lacking a coat and therefore rough in appearance (R). The S-type is highly infective when injected into mice, whereas the R-type is harmless, apparently made so because it has no protective coat against the body's defenses. The S-type bacteria have three variants depending on the kind of carbohydrate coat, designated I, II, and III. Griffith killed some S-type III bacteria with heat, and injected the dead bacteria into mice along with some living, but harmless, R-type bacteria that had mutated from a different S-type. Surprisingly, all of the mice died. Even more surprisingly, the dead mice were found to contain living S-type III bacteria! Griffith went on to show that the transformation from innocuous R-type to virulent S-type bacteria could be inherited.

Few, if any, bacteriologists believed the seemingly Jekyll–Hyde behavior of the pneumococcus bacteria, and Griffith was too shy to publicize his findings. But the bacterial transformation was confirmed by German scientists in 1928. Two years later a Canadian–American physician, Oswald Avery, at the Rockefeller Institute, now Rockefeller University, started working on the pneumococci. Avery and his co-workers were able to achieve transformation from living R-type to living S-type in a glass dish, instead of in mice, by culturing the R-type bacteria with a filtered extract from killed S-type bacteria. The transforming principle in the extract was effective at a concentration of one part in 600 million. Everyone thought that the mysterious S-type extract must be an enzyme (a protein) that changed the physiology of the R strain. But Avery and his associates found after lengthy and repeated experiments that the substance was pure DNA. Their results were reported in 1944.

Avery's finding upset the scientific apple cart because until then it had been thought that DNA was an insignificant cell constituent and that protein was the true genetic material. But it became apparent from Avery's work that DNA was the real genetic material and that it was being transferred somehow from the S bacteria to R bacteria and changing them genetically. Many people thought that Avery's discovery of the role of DNA justified a Nobel Prize, but Avery was too cautious in his conclusions for his discovery to have the impact that could have given him more acclaim.

Griffith and Avery demonstrated that genes can be transferred from dead bacteria to live bacteria by placing them where they could come in contact with each other. Is there really sex after death? Probably no microbiologist would put it that way, but philosophically, a point can be made for it. The essence of sex is the transfer of genetic material from one organism to another in a way that transmits some of the genes to the progeny. Nature, through evolution, has found innumerable ways of doing it. In many cases, there is no contact between the donor (male) and the recipient (female), although typically there are DNA couriers in the form of sperm and eggs. But the work of Griffith, Avery, and others later showed that genetic material of bacteria can be transferred to other organisms even after the death of the donor.

Japanese scientists made a related discovery in 1959 while working on inherited resistance. They found that dysentery-causing bacteria could pass their genes for resistance to *E. coli*, the common intestinal bacterium, and that *E. coli* in turn could pass the genes to *Salmonella*, still another type of bacteria, a troublesome intestinal pathogen. And in 1970, Robert Ferone and his co-workers, using the antimalarial drug pyrimethamine, were able to show that the malaria parasite *Plasmodium* could transfer drug resistance from one species of the protozoan to another. It is now commonly seen that genetic material is not only transferred readily from resistant to nonresistant strains of microorganisms, but to other species as well. Gonorrhea bacteria that are resistant to penicillin have a gene for producing an enzyme, penicillinase (now called β-lactamase), that breaks down the antibiotic before it can do its work. B. I. Eisenstein and co-workers at the University of North Carolina School of Medicine performed experiments showing that resistant gonococci can sexually transfer the penicillinase gene to other species of bacteria as well as among themselves.

Numerous pathogenic microbes have become drug resistant. They include staphylococci; *Streptococcus pneumoniae*; *Streptococcus pyogenes*; *Haemophilus influenzae*, a meningitis pathogen; *Neisseria meningitidis*, another important cause of meningitis; *Enterococcus faecalis*, responsible for 90% of endocarditis and urinary tract, wound, intra-abdominal, and pelvic infections around the world; *Vibrio cholera*; *Neisseria gonorrhoeae*; and the enteric pathogens *Shigella*, a major cause of dysentery, *Salmonella*, and *Campylobacter*, virulent causative agents of diarrhea.

Normally harmless, *E. coli* strains exist that cause serious infections, especially in hospitals and convalescent homes where other drug-resistant pathogens have also become increasingly threatening. Enterococci are now the third most common hospital-acquired organisms in the United States. Bacteria have become resistant to a wide range of antimicrobial drugs, including antibiotics, sulfonamides, disinfectants, and the metal-containing compounds of cadmium and mercury.

With all this microbial activity taking place, it should not be surprising that multidrug resistance has become common. Infection by a multiple-resistant pathogen can be both dangerous and costly. Dr. Harold Neu relates that he had to prescribe $25,000 worth of antibiotics to another physician who had contracted a strain of multidrug-resistant tuberculosis from a patient. Many people are unaware of the fact that tuberculosis is the largest cause of death in the world today, greater than cardiovascular disease or cancer, and that multidrug resistance to antibiotics by *Mycobacterium tuberculosis* is a growing threat.

Shigella, an organism transmitted by the fecal–oral route, is an example of one aspect of the problem. Although *Shigella* usually does not require drug therapy, antimicrobial therapy is given more often than necessary. Resistance appears rapidly by the transfer of plasmids, small circular elements within cells that contain genetic material outside the chromosomes.

Viruses are among the peskiest of pathogens, and their proclivity for shuffling genes makes them tricky and elusive targets. Viruses are not in the mainstream of the evolution of life. Whether they are living organisms depends on your definition of life. One view is that they are renegade DNA and RNA that escaped from respectable cells and went out on their own. The name *virus* was coined in 1898 by a Dutch botanist, Martinus Beijerinck, who filtered the juice of tobacco plants to remove any possible bacterial cause of the tobacco mosaic disease, and called the clear infective agent a *filterable virus*. *Virus* is Latin for poison. Further information that something unusual was going on in the submicroscopic world became known in 1917 when the French–Canadian Felix D'Herelle, while working at the Pasteur Institute in Paris, noticed spots in his cultures where there were no bacteria. He became convinced that something was killing them, and called the bacteria killer a *bacteriophage*, meaning *bacteria eater*. Such bacteria killers are now commonly called *phages*, and are known to be viruses. Phages had been discovered earlier by an English bacteriologist, Frederick Twort, but he did not follow up on it.

When Wendell Stanley, working at the Rockefeller Institute for Medical Research, now Rockefeller University, managed in 1935 to crystallize the tobacco mosaic virus, it was realized that viruses had some other remarkable properties. When in the crystalline state, they are as lifeless as a grain of salt. Some of them can survive in the environment for long periods, but they can proliferate only with the help of living organisms. They are surprisingly adept at transferring genetic material from one to another in a way comparable to sex. Electron microscope pictures show that viruses differ in size and shape. One of the first to be observed by electron microscopy has a hexagonal head and tail. The head consists of DNA and RNA with a wrapper of protein. They reproduce by injecting their genetic material into living cells and using the cells' metabolic machinery to produce more viruses.

Influenza is a classic example of how a virus can challenge the most sophisticated medical assault on a disease. A new virus may appear either as the result of simple genetic mutations, called *point mutations*, or by a process of gene transfer that is analogous to sexual reproduction, sometimes referred to as *genetic reassortment*. The outcome is a virus against which there is little or no immunity, and one that may cause a pandemic of influenza such as the Asian flu in 1957 and the Hong Kong flu in 1968 and 1969. Virologists speculate that the viruses of those epidemics resulted from human viruses that acquired one or more genes from avian (bird) influenza viruses during co-infection by the two viruses in the same animal or human. Because the virus is capable of so readily mutating and passing its genes around, influenza is called *a moving target*. The acquired immunodeficiency syndrome (AIDS) virus and similar viruses may turn out to be as evasive.

The sex imperative in the small world of microbes means a big world of trouble for humans.

Chapter 4

The Gametes

We've seen that unicellular species learned in the early dawn of life that sex gave them a leg up on survival in a changing world. At first they did it by clinging together while transferring genetic material. But it was not long, perhaps only a few million years, before they found a more efficient way of doing it. The new way was to use couriers—gametes—to carry on the work of transferring genes to their progeny. The new way proved to be the most durable innovation in the history of life since the invention of sex. The old familiar way of reshuffling their genes accomplished the purpose, but it was clumsy and often wasteful of precious cytoplasm. They did not precipitately switch to the new technique. Many of them continued with the old way for a long time, even after taking on the new.

Gametes can be thought of as distinct living organisms in the life cycle of sexual animals. One kind of gamete—the sperm—delivers the DNA to another kind of gamete—the egg, or ovum—that receives the sperm's DNA and fuses it with its own DNA. The union forms a fertilized egg, or *zygote*, that sooner or later proceeds to develop into an individual of a new generation. Except in the most primitive form of sexual reproduction, there is a difference in size between the two gametes. The smaller gamete that delivers the DNA is a *spermatozoon*, or *sperm* for short, and the larger gamete that receives the DNA is the *ovum* or *egg*.

A primitive form of sexual reproduction in which gametes come into play can be seen in a unicellular organism of the genus *Pando-*

rina. It lives in colonies of 16 cells imbedded in jelly. The cells reproduce either by fission or by producing gametes of two sizes. The smaller one fuses with the larger one to form a zygote—equivalent to a fertilized egg. The difference in size between the two gametes foreshadows what is seen in higher animals, where the smaller gamete is a sperm and the larger gamete is the egg. *Pandorina* belongs to a group of organisms, closely related to algae, that appears to be a point of divergence leading to either plants or animals. Some biologists consider all nucleated unicellular organisms a separate kingdom called Protista. In that classification *Pandorina* is one of the Protista.

A further increase in complexity in the evolution of sex can be seen in *Volvox globator*, a little organism in the twilight zone between unicellular and multicellular organisms. *Volvox* is a distant echo of the duality seen in the sex life of more advanced multicellular organisms. Zoologists call it a colonial protozoan (a single-celled animal) because it is motile, but botanists call it an alga because some of the cells in the colony contain chlorophyll. Without arguing the point, it typifies a step in the evolution of sex. The organism is able to reproduce with or without sex, the old way or the new way.

Volvox lives in ponds where it exists in a ball-like spherical colony held together by a matrix of gelatinous material. Each cell in the colony is called a *zoid*. The colony of zoids forms a green jelly ball, not more than 1 millimeter ($\frac{1}{25}$ inch) in diameter, containing as many as 50,000 zoids imbedded in the sticky surface. Most of the zoids are specialized for nutrition and locomotion. They have red eyespots, chromatophore inclusions, and usually two whiplike flagella with which they flail the water in coordinated action, causing the ball to roll over and over in spirited gyrations that are fascinating to watch under a microscope. The zoids are connected by strands of cytoplasm, making the colony, in effect, a multicellular organism, a member of the Metazoa.

Other cells are specialized for reproduction. They divide by simple fission without being fertilized, producing great numbers of cells that form new colonies. Cells destined for sexual reproduction grow larger and can be recognized as female cells called *macro-*

gametes or male cells called *microgametes*, which will form sperm. The macrogametes are fewer in number, and are larger, being loaded with food material for nourishment of the new colony they will eventually start. The male cells divide repeatedly to form spherical bundles of tiny flagellated sperm, which will leave the colony and swim about in the water seeking mature ova. When a sperm finds and enters an egg, resulting in a *zygote*, nothing may happen immediately. But when conditions are favorable, sometimes long after fertilization, the zygote starts to divide and continues to do so repeatedly until there is a community of cells ready to start another colony.

The distinction between egg and sperm is more pronounced in the malaria parasite, *Plasmodium vivax*. The complexity of the life cycle of this unicellular organism indicates a long evolutionary relationship with its victims—humans and several species of mosquitos belonging to the genus *Anopheles*. Whether the parasite makes the mosquitos sick is speculative, but the duo is historically one of the worst killers of mankind and remains so today.

When a female mosquito bites a human victim (male mosquitos do not bite), she injects a small amount of saliva, which contains a toxin that is presumed to be an anticoagulant. If the mosquito is infected, the saliva carries with it some parasite cells in the form of spindle-shaped *sporozoites* that find their way through the blood stream to the victim's liver where they penetrate and live at the expense of the liver cells. Each parasite cell divides by multiple fission into a large number of cells called *merozoites*, which penetrate more liver cells. Some of the merozoites leave the liver and enter red blood cells where they feed on the red corpuscles and develop in stages into what are called *schizonts*, which by multiple division form 15 to 24 merozoites each. Some of the new merozoites attack more red blood corpuscles and form more schizonts. The cycle continues repeatedly, causing an attack of fever in the victim every 48 hours. Still other merozoites become male and female sexual cells called *gametocytes*. Sexual activity of the parasite can go no further in the human blood. It will have to wait for a mosquito to suck in some gametocytes along with its blood meal where the lower temperature in the mosquito's gut

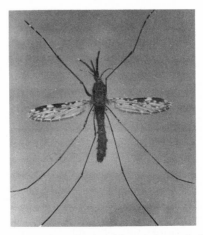

Human blood is too hot for sexual reproduction of the malaria parasite *Plasmodium* to go to completion. When the parasites are taken into a mosquito's stomach along with its blood meal, the cooler temperature allows gametes and zygotes to form. This is an adult female *Anopheles albimanus*, a tropical carrier of malaria in central America.

will permit developments to proceed. Inside the mosquito, the female gametocyte matures and becomes a *macrogamete*, comparable to an egg. The male gametocyte produces eight slender flagellalike cells called *microgametes*, comparable to sperm. After fertilization, the zygote changes into a wormlike cell that penetrates into the wall of the mosquito's stomach where it becomes an *oocyst*, producing hundreds of sporozoites. These break into the mosquito's body cavity, find their way to the salivary glands, and the circle is closed. The sporozoites are ready to be injected into the next human victim.

Receivers—The Eggs

Eggs, also variously called ova (singular, ovum), oospheres, or macrogametes, are the quintessential symbols of motherhood and regeneration of life. As repositories for the genetic material from the mother-to-be, the eggs are the primary vehicle for perpetuating the species and the nearest thing to making immortality of the living line theoretically possible.

Eggs of different species vary greatly in size and composition, depending mainly on the amount of nutrient material or yolk. They range from a few microns (micrometers) in diameter, as in many mammals, to several centimeters or several inches in diameter, as in some of the birds. Some eggs contain a supply of albumen, the egg white that serves as both nutrient and protective coating. The egg covering differs greatly among species. Some eggs are naked, having no coat other than the plasma membrane of ordinary cells. Naked eggs are common in some of the simpler organisms. Other eggs have a covering consisting of a proteinaceous or mucoproteinaceous membrane, as in some amphibians, some fish, and placental mammals. Others have chitinous cases, as in insects; keratinous envelopes, as in many reptiles; or calcareous shells, as in tortoises and birds.

The essential content of eggs is the genetic material derived from the parent. During formation of the egg, called *oogenesis*, a series of cell divisions culminates in an egg's receiving one member of each pair of genes that was originally the mother's. This takes place during two successive cell divisions called *meiosis*. Cells of the human body contain 46 chromosomes (23 pairs or two full sets of chromosomes) a condition *diploid*. During meiosis, cells are formed that contain only one chromosome of each pair, or only 23 chromosomes in all, a condition called *haploid*. The egg is formed by further divisions of the haploid cells. Eggs receiving chromosomes from the haploid sperm at fertilization will become *diploid* again with the full complement of chromosomes and the full complement of alleles being restored.

Searchers—The Sperm

Spermatogenesis—the production of sperm—is comparable to oogenesis in the reduction of genetic material to the haploid condition.

But there is an important difference between the genetic content of eggs and sperm. In many kinds of animals, including humans, gender is determined by chromosomes called X and Y. In mammals, for example, every egg carries a sex chromosome called an X chromosome, whereas half the sperm carry an X chromosome and half of them carry a Y chromosome. On fertilization, union between an egg (X) and a sperm (X) gives an XX zygote that will develop into a female, whereas union of an egg (X) with a sperm (Y) gives an XY zygote that will develop into a male.

Spermatozoa, or sperm for short, vary in size and shape among the various species. The difference in size between eggs and sperm is so great that, the variations among the tiny sperm pale in significance. Although mature eggs differ greatly in bulk, sperm differ relatively little in size regardless of the size of the animal. The sperm of a whale is no bigger than that of a mouse. This is not surprising in view of the function of sperm—to carry DNA, which is far from a bulky material. Sperm typically have a small head and a long whiplike tail, although the tails differ in size and length. The sperm of some toads are as much as 40 times the length of a human sperm, reaching up to 2 millimeters (0.08 inch).

Sperm of different species are identifiable by their distinctive shapes. The head of a frog's sperm is spindle shaped, and that of *Bufo* the toad is similar but with a sharp point and a ribbonlike undulating tail. The head of a turtle's sperm is like an elongated teardrop with a sharp point, and that of the robin has a long pointed head with a large filament spiraling around and down the length of the tail. The head of the salamander's sperm is long and slender, almost threadlike, with a filament spiraling around a long tail, and the head of a sturgeon's sperm is cylindrical with a blunt end.

The sperm of a crustacean has an unusual shape that defies verbal description, having a large head on a stalk followed by a bulbous structure from the base of which emerge several projections instead of a tail. That of the crayfish has a head like a blob of protoplasm with no tail, but instead, five short prongs projecting radially giving it the appearance of a pinwheel. Sperm of the worm *Ascaris* is conical in shape with no tail.

Mammalian sperm also differ greatly from species to species. The rat sperm has a hoodlike head, whereas that of the guinea pig has a large bulbous cap at the tip of the head. The sperm of the Chinese hamster has a tail several times as long as the human sperm. The sperm of the bull, dog, cat, and rabbit superficially resemble that of humans, although they are distinguishable by their size and shape.

For detailed anatomy of a sperm, the human sperm can be taken as representative. The mature human sperm is a single microscopic cell enclosed within a membrane, with a flattened head and a long, whiplike tail. Its volume is 85,000 times less than that of a human egg, which itself is barely visible to the naked eye. The overall length including tail is about 60 micrometers (microns), or about 0.002 inch. The head is four to five micrometers long, while the tail is about 12 or 13 times as long as the head. The thickness of the tail decreases gradually along its length. Sperm containing X and Y chromosomes can be easily identified by staining with the fluorescent dye, quinacrine, which causes the Y chromosome to fluoresce intensely. The staining technique works with human and gorilla sperm but not with bull, rabbit, or mouse sperm. Unfortunately, the technique cannot be used for controlling the sex of offspring because the staining procedure kills the sperm.

The existence of sperm was unknown until about 300 years ago. Their discovery had to wait for the invention of the microscope. In 1679 in the Dutch town of Delft, a few miles from Rotterdam, a 47-year-old drapery merchant and amateur scientist named Anton von Leeuwenhoek leaned over a primitive microscope he had made and peered excitedly at the squirming "animalcules" in a sample of semen. The sample was set up and previously examined by his student assistant, Johamm Hamm. When he later described what he had seen in a letter to the Royal Society, he was hesitant and embarrassed because he feared that it might appear to be obscene to the dignified and probably prudish members of the scientific society. But they took to him kindly and eventually made him a member, though he could neither read nor write Latin, which everyone used as the scientific language of the time.

There had been no more than a vague idea of fertilization when Aristotle speculated that the semen contained the "seed" that he thought of as growing like a plant and deriving nourishment from the woman. The word *semen* is from the Latin *sero*, meaning *to sow*. Some primitive people had little understanding of the relationship between sexual intercourse and babies. The Cro-Magnons, who made numerous figurines of female sex symbols, or so-called Venuses, seemed little taken with the male phallic symbol. But later people of ancient times saw the sexual connection clearly enough, and many were addicted to portrayal of the phallus and the sex act in various art forms and annual fertility rites, often with explicit performances by the celebrants. According to Robert Ardrey, the idea of fertilization never occurred to the Australian Aborigines or the Tobrianders.

The real nature of the sexual connection was not known until little more 100 years ago when in 1875 a German anatomist, Oscar Hertwig, detected the fusion of the nuclei of sperm and egg. This led to the understanding that fertilization and fusion of two sex cells brought together the qualities of the male and the female. It followed that the sex cell contains in its nucleus a hereditary contribution of the parent.

The life of a human sperm, as well as that of an egg, begins with primitive sex cells, called *primordial germ cells*, that form in the developing embryo long before any sex organs exist. They appear as early as 3 weeks after the fertilized egg begins to divide. As the tiny embryo grows into a blastula—a hollow fluid-filled ball of cells— then into a gastrula—the stage when the framework for the internal organs begins to take shape—the primitive sex cells migrate through the tissues of the embryo to where the kidneys are beginning to form, then on to the formative gonads, which at this stage are merely gonadal ridges.

All the while, the germ cells are dividing. Once they find their place in the embryo, they increase greatly in numbers. At first there is no detectable difference between the gonads of male and female embryos except by microscopic examination of the chromosomes. If the embryo is a male, the testes begin to form earlier than the ovaries form in a female embryo. Meanwhile, the germ cells are dividing at a great rate. They will increase to 600,000 during the second month when a human embryo has grown to 16–25 mm in length, and to almost 7 million by the fifth month. At this time the male germ cells are called spermatogonia, ordinary looking cells in the seminal tubules. Many of them will divide, forming two cells called primary spermatocytes, and the stage is set for the production of sperm.

Each primary spermatocyte divides, and at this point there is a reduction in the number of chromosomes in the process of meiosis. The chromosome pairs separate, one member of each pair going to the daughter cells, called secondary spermatocytes. These divide, each forming two spermatids that develop directly into sperm without further division. All this, from spermatogonia to sperm, takes place in the lining of the seminal (seminiferous) tubules of the testicles.

Meanwhile, spermatogonia continue to divide, resulting in an almost endless supply of sperm throughout a man's reproductive life. This is in sharp contrast to the situation in a woman, in whom neither the oogonia nor the oocytes (comparable to spermatogonia and spermatocytes) can increase their own numbers. In females of humans and most other species, the oogonia are usually used up by birth. The production of eggs declines with increasing age, and the supply is gradually depleted. Spermatogonia, on the other hand, begin to appear in the male infant's testes about 2 months after birth, and their formation continues throughout childhood. The testicles of the male child enlarge slowly after birth, and at the beginning of puberty there is a great increase in the number of spermatogonia. At maturity when the seminiferous tubules form, sperm production begins and may continue well into old age.

Spermatogenesis is rhythmic because all cells in a local area of germ cell tissue divide synchronously at regular intervals. The interval depends on the species. In the rat it is every 12 or 13 days, in the mouse 8.6 days, and in a man about 16 days. But there is a great deal of overlapping in tissue synchrony, so all stages are present at any given time, resulting in a production line that puts out a more or less steady stream of sperm at the rate of several million a day. Newcomers are always being packed into the vas deferens, so something has to be done to relieve the congestion, a situation that may provoke agonizing physical and psychological stress in the male. If there is no ejaculation by copulation, masturbation, or spontaneous emission such as commonly happens in youths having "wet dreams," the sperm will slowly leak out of the vas deferens into the urethra and be washed way in the urine. If the vas deferens has been severed by vasectomy, the sperm simply deteriorate, and the disintegration products are absorbed back into the tissues.

Insemination

During ejaculation, muscular contractions along the entire male genital tract from the epididymis onward force the sperm through the urethra and out the end of the penis. By this time the urethra has been moistened by fluid from two pea-sized Cowper's (bulbourethral) glands that serves to lubricate the penis in the vagina. As ejaculation takes place, simultaneous contractions pump a copious quantity of milky fluid from the prostate gland, which mixes with the sperm, making the semen slightly alkaline. The alkalinity is necessary for the sperm to become active and to counteract any slight acidity of the vagina. Sterility can result if the female genital tract is too acid for the prostrate fluid to neutralize it. The seminal vesicles contribute a secretion containing fructose, a sugar, that provides the sperm with a shot of quick energy. Semen from an ejaculation may contain an average of as many as 500 million sperm. If there are fewer than 60 thousand per cubic centimeter, the man will probably be sterile unless a special technique of artificial insemination is used.

When the human sperm finally reaches the uterus—if it gets that far—its life expectancy is something like 24 hours, during which it must either find and penetrate an egg or expire. Unless ovulation (release of an ovum from the ovary) is taking place, it is out of luck. The immediate goal is one of the oviducts, named fallopian tubes in honor of Gabriel Fallopius, a sixteenth-century priest-anatomist of Padua, Italy, who described the tubes. There is a chance that a few of the several million sperm deposited in the vagina will find their way through the uterus, into the correct oviduct, and make contact with the egg.

The tail of the sperm acts as a flagellum, lashing about violently in the fluid of the semen. Does this cause it to find an egg? From where the semen is apt to be deposited—in the vagina near the cervix of the uterus—the distance to where it might find the egg is something like a thousand times the length of the sperm. Tests on 24 ejaculates of bull semen from four bulls gave an average velocity of 94 μm (microns) per second (nearly $\frac{1}{4}$ inch per minute). Assuming that a human sperm could hold to a pace of 60 μm (its own length) per second, its speed would be fast enough to intercept an egg easily if it made a bee-line for the egg's position in the oviduct.

But that is not the way a sperm swims. It yaws and rolls, usually in a circle or curving arc, and seems to have little or no sense of direction. But in the end, the sperm probably does not have to depend on its swimming ability. Instead, it appears to be propelled, first, by the pumping action of the penis, then carried along by muscular contractions in the wall of the female genital tract during and after orgasm, which may continue for some time. The semen contains hormones called prostaglandins that stimulate muscular contractions, and the squirming around of the sperm probably helps to counteract friction against the wall of the genital tract. Sperm that get into the uterus still have to find their way to the oviduct, and even there they may find another unexpected barrier. It is estimated that half of the female-related sterility problems are caused by blockage of the oviduct in a way that prevents the egg from descending. By this time, attrition of the spermatozoan troops has been colossal. Many of the sperm go up the wrong oviduct, and because the ovaries usually alternate in releasing an egg, the wrong-way sperm reach a dead end. Out of several hundred million sperm deposited in the vagina, only a few hundred may reach the oviduct. But once in the oviduct, if everything goes well for the sperm, they are carried along rapidly, aided by muscular contraction and the beating of cilia in the wall of the oviduct. When they reach the ovum, one of them will be the winner.

But it is not a sure thing. The ovum is a tough egg to crack. A thick covering called the *zona pellucida* is enclosed in still another membrane that would be an impenetrable barrier if the sperm did not have a way of dissolving it. Unless they first undergo a physiological change in the uterus, a process called *capacitation*, the sperm will not be capable of fertilizing an egg because they will not be able to penetrate the egg's thick coat. The top of the sperm's head is encased in a cap called an *acrosome*, from the Greek *akros* meaning *tip* and *soma* meaning *body*. When the sperm bumps into the egg, the encounter stimulates what is known as the *acrosome reaction* whereby pores open up in the membrane of the cap, releasing enzyme that dissolve the jelly in the outer coat of the egg. In humans, normally only one sperm will fertilize the egg because changes quickly take place in the egg membrane that "close the door" and prevent other sperm from penetrating far enough to fuse with the egg's nucleus.

A sperm faces astronomically unfavorable odds. All stages of sperm development are susceptible to injury from infectious diseases, local inflammation, drugs, chemicals, alcohol, x rays, and even dietary deficiencies. Sperm are afflicted with various kinds of abnormalities. When more than one fourth of a man's sperm are abnormal, he will probably be infertile. High temperatures prevent sperm production because the germ cells in the testicles cannot divide unless the temperature is below body heat. Ordinarily the problem is taken care of by the scrotum—the bag holding the testicles—which serves as an air-conditioning device keeping the testicles cool enough to function, several degrees cooler than the abdominal cavity. In some mammals the testes remain in the abdomen during the part of the year when the animal is sexually inactive and descend during the breeding season.

If the scrotum is kept covered to raise the temperature, sperm will either fail to mature or be malformed. Occasionally during fetal development, a testicle will fail to descend into the scrotum, a condition called *cryptorchidism*, in which case viable sperm cannot be produced. If both testicles fail to descend, the male will be sterile; however, if attention is received soon enough, the condition can be corrected by surgery. It is speculated that the hot baths favored by the Roman patricians may have contributed to the decline in their offspring, but they had learned how to use hot compresses as a form of birth control. Some people wonder if prolonged exposure in saunas by Scandinavians and the lengthy hot baths of the Japanese have the effect of lowering their reproductive rate. Tight clothing, like jockey shorts, probably under some conditions inhibit sperm viability, whereas Scottish kilts and loose robes like those worn in the Middle East and parts of the Far East probably favor sperm survival.

Sperm and egg became nearly the universal vehicle to transfer genetic material from parents to their progeny. The gametes have continued to perform the function of fertilization from protozoa to primates under changing conditions throughout evolutionary history. The way organisms bring them together depends largely on the environment, especially whether wet or dry.

Part II

Connections

Chapter 5

The Wet Connection

Sexual reproduction was invented by organisms that lived in the sea or in brackish or fresh water. When conjugation and fusion were fashionable, the watery environment was no problem. But when gametes came into being, allowing the advantages of greater mobility and fecundity, the vagary of currents and other hazards called for new strategies. One of the best ways to have enough progeny for the species to survive was to produce a redundancy of eggs and sperm. This is common in marine animals. A cod will lay up to 8 million eggs, though usually fewer, which float on the surface of the sea. A Pismo clam produces about 15 million eggs a year. There could be 100 trillion eggs spawned by the clams along less than 10 miles of beach, and if they all matured into legal-size clams and were laid end to end, they would encircle the earth 300,000 times. But only a small fraction of a percent of the eggs ever become mature clams. An annual census showed that in 1 year, only 33,000 clams resulted from 120 trillion eggs.

When the males of aquatic species such as fishes release their milt there is often a visible cloud of millions of sperm. This early strategy carried over into more evolutionarily advanced animals and may account for the apparent redundancy of sperm even in terrestrial animals where contact is intimate and there is a minimum of waste.

An example of simple watery sex life is seen in the sea anemone, *Metridium*, which has separate males and females. They shed their gametes into their large gastrovascular cavities and expel them through their mouths to the outside water. The fertilized eggs develop into free-swimming *planula*, which find a place to attach themselves on rocks or other solid material where they develop into adult sea anemones.

Animals that lived in the sea devised numerous strategies to overcome the risk of their gametes failing to find each other by simply floating or swimming around in the shifting water. The lowly sponges developed a clever innovation. Sponges have been known since ancient times but it was not until 1857 that they were recognized as animals. There is dispute about where they fit in the evolutionary scheme, but one view is that they are in the mainstream of the evolution of the Metazoa (multicellular animals). *Scypha* is a small sponge, not over 2 cm long, shaped like a slender vase that bulges in the middle. It can be seen along the eastern and western seacoasts of the United States attached to rocks near the low-tide line. In its early embryonic stages it has external flagella, tiny whiplike organelles that enable it to swim around before settling down. It is *hermaphroditic*, that is, it can reproduce asexually or sexually. It reproduces asexually by forming a bud near its base, which eventually becomes free and forms a new sponge. Sexually, it develops separate structures that produce eggs and sperm in the same individual. A structure called an oogonium divides twice to form four eggs, while another structure called a spermagonium divides to form a cell called the sperm mother cell, which divides to form several so-called spermatids that develop directly into sperm. In a strategy of conservation, the sperm are not released in the water to swim around looking for an egg. Instead, they become attached to special cells called *choanocytes*, which are cells modified to become amoeboid and carry the sperm to the eggs before they are released.

One of the most primitive small worms, the fresh-water flat-worm, *Dugesia tigrina*, a common planarian, has an extremely flat, unsegmented body that may reach a length of 2 cm. Planarians are noted for their remarkable powers of regeneration. When cut in two, the head end will regenerate a new tail, and the tail end will regenerate a new head. Even when cut into smaller pieces, each piece not below a minimum size will regenerate an entirely new individual. An equally remarkable quality is that when planarians are trained to have a simple conditioned response using an electric shock, then chopped up and fed to untrained planarians, the untrained individuals that feed on the minced intellectuals are found to have acquired the same conditioned response. RNA is the memory-transferring substance.

Planarians have a remarkably well developed, but limited, organ system. They have muscular, digestive, excretory, reproductive, and nervous systems, including a brain and two eyes, but there are no circulatory or respiratory systems and no anus. Each individual has a full set of both male and female reproductive organs. There are several grapelike clusters of testes connected by tubes to a penis, a number of prostate glands, a vas deferens, and a seminal vesicle where the sperm are stored. There are two ovaries farther toward the front end of the body, each with an oviduct, yolk glands, a vagina that opens into the genital cloaca, and a uterus.

The planarians are a classic example of the biological imperative of cross-fertilization. Although hermaphroditic, the genital organs are arranged so that self-fertilization is impossible. When they copulate, each of them uses its penis to transfer sperm into the other's genital cloaca. Fertilization occurs in the uterus where cocoons are formed that contain from 4 to 20 eggs surrounded by yolk cells. The penis in some of the flatworms is armed with knifelike stylets that are used to puncture the body wall of the other worm, enabling the donor to inject sperm into the body cavity by hypodermic insemination.

The tapeworm, *Taenia solium*, is one of the most prolific reproductive machines ever invented by nature. Unlike the planarian, it has escaped from its presumed aquatic ancestry to become an intestinal parasite. And instead of being a simple flatworm, it has a head with a short neck followed by a string of up to several hundred flat segments, called *proglottids*, each a complete sex machine in itself. Each proglottid contains both male and female reproductive systems: numerous grapelike testes attached by tubules to a vas deferens ending in a genital pore, two ovaries attached to an oviduct, a uterus, a vagina, yolk glands, and a shell gland. The male and female systems merge at the single genital pore. Unlike planarians, each proglottid can fertilize its own eggs. Mature proglottids at the end of the string eventually break off and are passed out with the feces.

With as many as 800 to 900 proglottids reaching a length of as much as 3 meters, a single tapeworm can put out an enormous number of fertilized eggs. If the eggs are eaten by a pig, the embryos bore through the intestinal wall and find their way to voluntary muscle tissue where they encyst and develop into larvae called bladderworms. If infected pork is not thoroughly cooked, a person eating the meat is apt to be victimized by the intestinal parasite.

All flatworms are hermaphroditic except schistosomes, in which the males and females are separate. Schistosomes are blood flukes that attack birds and mammals. Three species are serious parasites of people, one of them causing a historic—and current—debilitating disease in Egypt where the intermediate host, a species of snail, inhabits the irrigation canals along the Nile River. The snails give rise to minute free-swimming *cercariae* that penetrate the skin and get into the blood of people who come in contact with infested water. The mature male is larger than the slender female and carries her in a deep groove that extends almost the length of his body. The cercariae of bird schistosomes try to penetrate the skin of humans, but they can only get far enough to cause an irritation known as *swimmer's itch*.

Advocates of *the selfish gene*, the sociological theory that the organism is a physical device that has come into existence to perpetuate the genes, can find support in the spinyheaded worms belonging to the phylum Acanthocephala. These worms are parasitic in vertebrate animals in which they become attached to the intestinal walls by a proboscis covered with hooks. The immature stages are found in arthropods such as insects and crustaceans. The mature male has large cement glands, with a receptacle to hold the cement, which he uses to plug the genital opening of the female after mating. The plug prevents

The female schistosome is longer and more slender than the male. She lives most of her life in a groove along the underside of the male. Schistosomes are blood parasites with complicated life histories. Three species cause serious diseases in parts of Africa and Asia. Drawing by Vicki Frazior.

loss of the sperm and blocks subsequent copulations by rival males that might be eager to introduce competing sperm. Some of the successful males go a step further. They render other males impotent by plugging their genital openings with cement during homosexual rape.

The squid, *Loligo pealei*, has a unique conservation strategy that minimizes loss of sperm during transfer to the female. A cephalopod related to the octopus, the squid has a streamlined spindle-shaped body with 10 appendages at the head end. It propels itself through the water by hydrojet propulsion using a so-called *funnel* through which water is squirted when it squeezes down with its covering muscular *mantle*. The squid uses the funnel as a steering device by directing the jet according to where it wants to go. The squid is notable for its large brain and two large, well-developed eyes with lenses and retinas.

Eight of the squid's head-end projections are called arms, and two of them, longer than the others, are called tentacles. Both arms and tentacles have numerous suckers on their inner surfaces. One of the arms, a so-called *hectocotylus*, meaning *hundred suckers*, has a sexual function. The male has a large testis from where sperm are transferred through a vas deferens to a convoluted structure called the spermatophoric organ, which compacts the sperm into long slender packages called *spermatophores*. These are stored in a spermatophoric sac until needed. At mating time, the packages of sperm are ejected by the squid's penis, picked up by the lower left arm— hectocotylus arm—and deposited safely into the female's mantle cavity. The action of sea water causes the spermatophore to release a sac filled with spermatozoa in a sudden *ejaculatory reaction*. The sac is open at one end from which it releases the sperm slowly over a period of hours, and they fertilize the eggs as they are extruded by the female. She produces a large number of eggs, which she coats with a secretion before they are laid, and deposits them on the bottom where they hatch directly into tiny squid. In the related argonaut, also called the paper nautilus, *Argonauta*, the hectocotylus breaks off and stays in the female. Two species of octopus also have a detachable hectocotylus. Early observers thought it was a parasitic worm in the female.

Squids have the ability to change color quickly from bluish white to mottled brown and red. Their coloration often blends with the background and serves as a camouflage. The skin, especially on the dorsal side, contains numerous elastic sacs filled with red and yellow pigments. Variations in the visibility of the pigments, and therefore color, are produced by relaxing and constricting muscles under nervous control. The display can be spectacular during the male and female sexual embrace.

Most of the fishes rely on releasing large numbers of eggs and sperm into the water and trusting to luck that enough eggs will survive and enough sperm will find the survivors to perpetuate the species, but various strategies can improve the odds. The adventurous life of the salmon is one of the most romantic stories in wildlife lore. Some species of trout and nearly all salmon spend a great part of their adult lives in the open sea but return to the rivers of their birth to spawn.

Salmon, especially, will travel great distances from their birthplace in the rocky bottom of a river near its headwaters and return to the same river of their birth after 1, 2, or 3 years of feeding in the ocean. They will persevere in the long journey upstream in a battle against currents, rapids, dams, and the uncertainties of fish ladders. They are guided by their sense of smell and taste. Each river has its own mineral and organic content and, no doubt, its own flavor.

The jaws of the salmon become long and hooked as breeding time approaches, and there are changes in color, especially pronounced in the male. Males and females pair up, and the male becomes aggressive, on the alert to fight off other males that threaten to approach his mate. The female makes a shallow trough in the gravel with her tail and lays her eggs in it. The hovering male sheds his sperm over the eggs, and the female promptly covers them with gravel. Sometimes a young male that has not yet gone to sea will maneuver close to the nest and shed his sperm, cuckolding the older male by fertilizing at least some of the couple's eggs. To add insult to injury, the young male, at this stage called a *parr*, is apt to cannibalize some of the eggs of the older couple, gaining strength before heading downstream to the sea with others of his generation. Sometimes a male trout will cuckold a male salmon. If he can get away with it when the male salmon is not looking, he will close in on the spawning female and shed his sperm over the nest. The hybrids from the cross between trout and salmon may hatch and develop, but the hybrid females have low fertility, and the males are completely sterile.

An innovative and efficient way to brood young fish is the habit of the female bitterling, *Rhodeus amarus*, of laying her eggs in the siphon of a swan mussel. If a male bitterling and a mussel are both present, the female's cloaca (the all-purpose genital–urinary–anal opening) develops into a long tube that functions as an ovipositor. Its growth is dependent on both her secretion of progesterone and a hormone released into the water by the male. When the female is ready to deposit an egg into the mussel, she passes the egg into the ovipositor, which becomes erect by the pressure of urine forced into it.

As she deposits her egg in the mussel, the hovering male sheds his milt close to the mussel's opening, and the sperm are swept in by the water that is sucked into the mussel's siphon. The female bitterling unknowingly reciprocates the favor. As she is laying her eggs, the mussel releases its larvae, which attach themselves to the fish's skin, scales grow over them, and they are nourished there until they break out as fully formed miniature mussels.

Some species of fish improve the survival rate of their offspring by brooding them in their mouths. In some catfish, the young develop in the mouth of either the female or the male. In the African mouth-brooder, *Haplochromis burtoni*, the female collects the eggs in her mouth as she lays them. The male comes close and as he sheds his milt the female, seeing bright spots on him that she mistakes for eggs, gulps in sperm as she tries to snap them up.

Guppies (*Lepistes*), mosquito fish (*Gambusia*), and a few others have developed a system of internal fertilization. The male has an intromittent organ called a *gonopodium*, a tube formed by a modification of the anal fins immediately behind the genital opening. The gonopodium is used to funnel the sperm into the female's genital opening.

Sea horses and pipe fishes belonging to the family Syngnathidae not only developed internal fertilization but defy convention by reversing their sex roles in a way that raises questions about the evolutionary significance of their structure and behavior. The sea horse, *Hippocampus ingens*, has a grotesque shape suggesting the head and neck of a horse and has a protective resemblance to the seaweed on which it lives and feeds. An elongated caudal fin serves as a prehensile tail that it wraps around the stems of the seaweed. These fish copulate upright in the water face-to-face. The male has a brooding pouch formed by folds along the belly. The female is larger and brighter than the male and takes a more aggressive role in courtship. She has an intromittent organ resembling a penis that she inserts into the male's brooding pouch and deposits her eggs there. There is a reversal in roles but no change in sex because the female, as usual, produces the eggs and the male produces the sperm. The male fertilizes the eggs and broods them until they hatch. In some species the male nourishes the young fry by means of tissue comparable to a placenta. Gestation takes about 3 weeks, and when the young are ready to go out on their own, father expels them in a cloud of fully formed miniature sea horses.

Why this reversal of sex roles? Usually the female has the greater investment in producing offspring. In the sea horse, the female has considerable investment because the eggs are heavily endowed with yolk, but she is able to improve her productivity by foisting part of the work onto the male. William Eberhard speculated that the reason males are usually the ones with intromittent organs is that if they have less invested in reproduction, they can afford to spread their genes around to more than one female, whereas the female has little to gain by copulating with additional males because it will not increase her productivity. The male, with his intromittent organ, is better qualified

The female sea horse, *Hippocampus ingens*, conserves her reproductive energy by switching roles with her mate. Using a penislike organ, she inserts her eggs into a pouch on the male's belly where they remain until he expels the brood as fully formed miniature seahorses. Redrawn by Vicki Frazior from Daniel J. Miller and Robert N. Lea, *Guide to the Coastal Marine Fishes of California*, Fish Bulletin 157, California Department of Fish and Game, 1972.

to compete with other males than if the roles were reversed. Natural selection favors those that win out in the competition for passing their genes to future generations.

Sharks, and other fishes that have skeletons of cartilage instead of bone, are all equipped for internal fertilization. In the male, the glands of the vas deferens produce a substance that causes the sperm to aggregate into spermatophores. The organ used for transfer of the spermatophores to the female consists of parts of the male's pelvic fins modified to form a large pair of *claspers*. The organ contains erectile tissue and a pumping mechanism. Erection is nerve dependent and appears to be activated by the secretion of epinephrine (adrenaline), because injection of the hormone produces an erection. During copulation, the claspers are inserted into the female's cloaca. Some sharks are viviparous, giving birth to living young. Others produce eggs enclosed in an elaborate shell called a *mermaid's purse*.

Crustaceans, a group of the arthropods (jointed-leg animals), are with few exceptions aquatic animals. They are represented by such creatures as shrimp, lobsters, crabs, and numerous smaller organisms, all of which have complex structures, peculiar adaptations, and unique life styles. Some of the crabs spend almost all of their lives on land but return to the water to reproduce. The genital pore of the male crayfish is in the fifth walking leg from the front, and that of the female is on the third walking leg. Copulation takes place during regular mating seasons. The male approaches the female and grabs her by her head appendages. She usually resists, but after a struggle he turns her over on her back and stands over her while transferring sperm into her genital pore. She will lay up to several hundred eggs in a complicated procedure during which she sticks them to the hairs of her swimmerets, where they are held safely until the larvae hatch.

Barnacles look superficially like mollusks, related to clams and mussels, but they are actually crustaceans protected by calcareous shells. The larvae have the habit of settling down and forming tightly packed encrustations on rocks, pilings, and boat bottoms, much to the annoyance of boaters. Unable to move, barnacles might be supposed to fertilize each other by releasing sperm into the water, but they have managed to avoid that costly gamble by using a long, slender penis that makes it possible to deliver sperm to nearby neighbors.

Frogs and toads developed some of the most remarkable and intriguing reproductive strategies in the animal kingdom. Along with other members of the 2000 or so species of Amphibia, they made giant steps in preparation for life on land. The name amphibia means literally *double life*, from the Greek words *amphi*, meaning *both*, and *bios*, meaning *life*. There is fossil evidence that they evolved from prehistoric lungfish that made brief and precarious excursions on land beginning around 250 million years ago. Amphibians acquired many of the features that are familiar in higher terrestrial animals—the first of the Tetrapoda (four limbs), and pentadactyl (five digits on each limb), and air-breathing in the adult stage. The legendary jumping ability of frogs and toads was immortalized by Mark Twain in his classic yarn, *The Celebrated Jumping Frog of Calaveras County*. And there is the still-popular comedy *Frogs* written by Aristophanes in 405 BC.

Amphibians superficially represent a link in the greatest achievement in the evolutionary history of animal life—the gigantic step that fishlike sea animals took when they left their watery cradle for the great adventure on land. But the amphibians, with few exceptions, never made their way completely out of water. To be sure, there are dry-land species like the desert toads of Australia that survive by burrowing and have physiological adaptations for conserving water, but most of the amphibians must have access to water. Their eggs are not equipped to withstand dryness. Frogs, in contrast to toads, have a smooth, thin, more or less slimy skin that is more susceptible to desiccation than that of toads, which typically have warty, drier skin that enables them to be more fully terrestrial.

Frogs and toads are notoriously lingering lovers. At mating time, the male is usually the first to return to a pond, stream, or similar breeding place, where he waits for a female to arrive in response to his call. When the female's eggs are ripe, she enters the water where the male mounts her piggyback fashion and embraces her by clasping his forelegs around her thorax and holds her so tightly around the breast that it is almost impossible to break his grip short of killing him. He is able to hold on with the help of *nuptial pads*, that form a sort of extra digit on the inner side of the front legs. One early experimenter, deficient in romantic sensitivity, could not induce a male to loosen his hold even by amputating his hind legs. Males of some species will hold on to the female for days or weeks without letting her go.

But what is going on is not what you might think. The male has no penis, so all he can do is discharge his sperm over the eggs when the female extrudes them. Females are far from choosey. If a wire clip or padded clothespin is clamped to her body in the same position as the male's forelegs would be, she will extrude her eggs as if she were being mounted by a male frog. The mating embrace is called *amplexus*. Except for the bell toad, *Ascaphus truei*, which lives in swift, cold mountain streams of the Pacific Northwest, and an African frog, *Nectophrynoides*, which gives birth to living young, there is no contact between the male and female reproductive organs. One exception is the male *Ascaphus*, which has a penislike extension of the cloaca that enables it to ensure internal fertilization. Also, the eggs of a group of primitive amphibians, commonly called caecilians, are fertilized internally. These wormlike, burrowing creatures have long, slender bodies with no limbs. They are carnivorous animals that feed on worms and other small subterranean life. The male copulates with the female by means of a protrusible cloaca that performs the function of a penis. During a sort of courtship ritual of salamanders, the male of some species extrudes a packet of sperm on a leaf or other object where the female picks it up with the lips of her cloaca.

Some male frogs have a sneaky way of having an amplexus with a female. Many male frogs and toads commonly give out loud croaky signals specific to their species to advertise their presence to females. But often in the same pond or puddle there are males of the same species that remain strangely silent. Stephen Perrill of Butler University and his co-workers made a study of the green tree frog, *Hyla cinerea*, to determine what the Silent Sams were up to, and found out that some of them would waylay females and grab them on their way to answering the croaker's call. Not all croakers took the intrusion lying down. They would often chase the spoilers away, or attack while giving pulsed *encounter* calls, and engage in butting and wrestling. Similar sexual freeloading has been reported in fish, iguanid lizards, elephant seals, and ruffs. Conservation of croaking energy seems to be a useful strategy when it works.

A female frog typically lays her eggs in a jelly-coated cluster, forming a tapiocalike mass that adheres to plant stems and other objects in the water. The toad lays her eggs out in a string resembling a gelatinous string of beads. Each jelly-coated egg becomes an embryo, which emerges after a few days from the jelly as a small larva, commonly called a tadpole or pollywog. If you look into a temporary pond in spring, you are apt to see it teeming with wriggling pollywogs. They start out with plump bodies, long tails, gills, and horny teeth to scrape algae off vegetation. As they develop, they absorb their tails, lose their gills, develop lungs, grow legs, and hop out of the water as miniature frogs or toads. There are corresponding changes in physiology and habits. Pollywogs are strict vegetarians, their diet consisting largely of algae, but when they become frogs or toads, they are carnivorous, eating insects or whatever prey is handy. A giant frog, *Rana goliath*, of West Africa grows more than a foot long and is said to eat animals as big as rats and ducks.

The number of progeny produced by frogs and toads varies. There is a small Cuban frog, *Sminthillus*, that lays only one egg, but other species lay up to thousands. The female bullfrog, *Rana catesbeiana*, will produce as many as 25,000 eggs, and the large poisonous toad, *Bufo marinus*, may lay up to 32,000 eggs at a time.

Frogs and toads display a bewildering variety of reproductive habits. Parental care is a prominent and surprisingly varied evolutionary development in many species of amphibians. Some amphibians make unusual nests for their young. A toad, *Leptodactylus mystacinus*, found in Texas, Mexico, and South America makes a bubble bath for the young to frolic in. She does this by working up a frothy mass of mucus in a cavity in the ground near a stream. The tadpoles are able to complete their development in the bubble bath where they remain until they grow big enough to hop their way to nearby water. Other froth-making species put the bubbly mass of mucus on a leaf that is then folded into a basket, forming a pocket that protects the young from predators during the early part of their growth.

The male of a Brazilian tree frog, *Hyla faber*, builds a little circular swimming pool formed by a ring of mud projecting above the water in a shallow pond. The female lays her eggs in this protected pool, which is about 12 inches in diameter and 4 to 6 inches deep. The tadpoles remain in their private pool until mature enough to hop over the barrier. A related tree frog, *Hyla resinfictrix*, builds the young-sters' private pool high in the air. It finds a cavity in a hollow tree and lines it with beeswax that it gets from the combs of certain species of stingless bees. When rainfall fills the cavity, the female lays her eggs in the water where the tadpoles can develop protected from enemies.

A small toad, *Nectophryoides occidentalis*, found near the summit of Mount Nimba in Africa, is born as a fully formed toad, having spent its larval life during 9 months of pregnancy in its mother's uterus. This species has no aquatic stage at all, and some of the typical tadpole features never appear or remain vestigial; the larva does not develop the tail fin, gills, spiracle, branchial cavity, or the horny teeth of other tadpoles. At about the eighth month of gestation, what tail fin it has disappears; it quickly forms limbs and is ready to come into the world as a mature frog.

Many of the tropical amphibians have strange breeding habits. The Surinam toad, *Pipa pipa*, of northern South America has a unique way of keeping the eggs and young under the protection of the parent. When breeding season comes, the skin of the female's back becomes pitted with cavities, forming small pouches in the skin supplied with a network of blood vessels for a rich supply of food. As the female lays her eggs, her oviduct protrudes like a bladder and is pushed over her back by the belly of the male. He manipulates his body in a way that spreads the eggs over the female's back and presses them into the small cavities. A membrane forms over each cavity, making a lid that completely encloses the eggs. The embryos go through their complete transformation to fully formed miniature toads within the cavities. When the young toads break out and leave their mother, they are ready to strike out on their own without ever having been in the water. They grow to be 5 inches long and have flattened bodies and long webbed toes.

A similar method of caring for the young is used by the marsupial frog, *Gastrotheca* (or *Nototrema*). The female has a large pouch on her back formed by a fold of skin that opens toward the rear. As the eggs are laid, the male shoves them into the pouch where the larvae develop before emerging as frogs. Another method of protecting the young is used by an Australian frog, *Assa darlingtoni*, also called the marsupial frog. After the female lays her eggs, the male hops into the egg mass and wriggles around until he becomes surrounded with it. When the tadpoles hatch they make their way through the gelatinous mass on the father until they find the opening to a sac located on his flank where they complete their development.

The female of another Australian frog, *Rheobatrachus silus*, has an innovative way of raising her young. She swallows the fertilized eggs and carries them in her stomach until they become fully formed juvenile frogs. The male of a Chilean tree frog, *Rhinoderma darwini*, takes over a similar role (rhinoderma means literally *rough skin*). It has completely escaped dependence on bodies of water. When the female lays her eggs, the male fertilizes them, then pushes them down his throat into his enlarged vocal sacs where he carries them until they develop into tiny, fully formed frogs ready to be burped out.

An otherwise nondescript European amphibian called the midwife toad, *Alytes obstetricans*, has an unusual reproductive life in that the toads mate on land. When the female extrudes her eggs in a long, necklacelike string of jelly, the male exudes his milt over them, then winds the strands of beads around his legs where he carries them until they hatch. Meanwhile, he stays in damp places so when the little tadpoles emerge from the eggs, they can drop into a convenient pool to complete their development. Thus he acts the part of a midwife, or obstetrician as the Latin name implies, tending the birth of the infant toads.

We have seen that following the innovation of gametes, numerous strategies were adopted to get sperm and eggs together. Fertilization of eggs by sperm in a water environment—where gametes were invented—runs the risk of wasting biological energy and resources if conditions do not favor contact between sperm and egg. Females have the greater investment to protect because nutrient material such as yolk is usually required for egg production, but the male also has a stake in getting the maximum productivity out of the sexual encounter. The function of both parents is to produce as many offspring as possible to perpetuate their genetic legacy in the battle of survival of the fittest. The variety of successful strategies to accomplish this gives no clue to the number of innovations tried or how many failed. That many of them succeeded is evident from the numerous sexually reproducing species that managed to survive and evolve in their aquatic environment. More innovations were needed, as we shall see, when animals ventured into the unknown environment of a terrestrial life.

Chapter 6

The Dry Connection

Life in the sea and in other bodies of water was the only life on earth for 2 billion years or more after the first living organisms appeared. For most of the earth's history, the exposed land was barren rock. There could be no animals where there were no plants. The conventional view is that the simple one-cell plants known as cyanobacteria (formerly called blue-green algae), which were abundant in the seas by at least 3.5 billion years ago, eventually got a foothold on land and slowly proliferated as some of the primary occupants. Then, about 500 million years ago, higher plants became established and took over. Animals crawled out of the sea to eat the plants, and eventually some of the animals learned to eat the plant-eaters. Higher plants and terrestrial animals were probably well established by 400 million years ago.

But the previously perceived time scale may have to be slightly modified. Paul Knauth, an isotope geochemist of Arizona State University and Robert Horodyski of Tulane University came up with evidence of microfossils, probably cyanobacteria, that might have been living on land several hundred million years before higher plants appeared. Cyanobacteria take in carbon dioxide as part of the process of photosynthesis. They favor the lighter of two stable isotopes of carbon, and when they die and decay, they release most of it in the form of carbon dioxide, which is absorbed by water and may be deposited as carbonate in the rock on which they live. Determination of the relative abundance of the two isotopes of carbon indicated that photosynthesizing organisms like cyanobacteria might have been present on land 1.2 billion years ago.

Before animals could become fully terrestrial, they abandoned the slipshod procedures of external fertilization by males and females releasing their gametes near each other in water in favor of the more reliable and usually less prodigal method of internal fertilization. They invented several ways of doing it. The terrestrial arthropods (jointed-leg animals), consisting of a million or more species, display a vast array of forms and life styles. One group of crawling, land-dwelling creatures, the Class Arachnoidea, includes spiders, scorpions, ticks, and mites. There is great variation among them in anatomy, physiology, and behavior. One of the more innovative reproductive devices is that of scorpions. They engage in a courtship ritual involving a sort of "dance." The male fastens a spermatophore to a rock or similar surface and maneuvers the female around until her genital opening is over it. The spermatophore has a pair of hooklike triggers that, when the female makes contact with them, cause the spermatophores to snap open in the middle with a force that shoots the spermlike projectiles into her genital opening.

Spiders have an entirely different approach to the problem. None of the spiders has true jaws or antennae. The mouth of the spider is only a tiny opening; it cannot take in solid food, so a spider subsists on the juices of its prey. The first appendage is a pair of *chelicerae*, which typically have structures at the base sometimes called mandibles, for smashing victims in preparation for sucking out their juices. The chelicerae have claws at the tips (the word chelicera is from the Greek *chela*, meaning *claw*) and poison glands. Poison pours out at the tips to kill captive insects. The venom is strong enough in some species to severely injure large animals, including humans. The second pair of appendages from the front are all-purpose *pedipalps*, which are used as jaws to hold and help macerate victims and for copulation. There are four pairs of walking legs, the bases of which sometimes also serve as jaws. Male and female sexes are separate. The female's oviduct and the male's vas deferens open to the outside at their respective genital pores on the underside near the front end of the abdomen. The female also has a seminal receptacle just inside the opening.

During copulation, the male spider typically smears his pedipalps with sperm taken from his genital pore and transfers it to the genital pore of the female. The eggs are fertilized within her body. She lays her eggs in a silk cocoon that is usually fastened to the web or some other object, but in some species, she carries it around with her. In most cases, the young emerge from the cocoon in a horde almost as soon as they hatch.

The practice of some female arthropods, especially insects and spiders, of murdering and eating their mates after copulation is a familiar story. The tale of the black widow spider, *Latrodectus mactans*, is one of the most lurid. The cannibalistic nature of this Lady Bluebeard of the web spinners is uncommon but understandable

The black widow, *Latrodectus mactans*, recycles the male by consuming him after mating. His body helps nourish the eggs.

as a strategy for survival of the species. Spiders are carnivorous, and the time of greatest need for nutrients is during reproduction. Her attack on the spent male cannot be a case of mistaken identity because spiders, usually having eight eyes, have excellent eyesight, probably acute to a distance of 12 centimeters (about 5 inches). The truth is that after fertilizing the eggs, the male black widow is useless, and his substance might as well be used to nourish the embryos and help perpetuate the species. Sometimes, though, an alert male can beat the odds and escape to live and love another day.

The male of the species, in contrast to the large, globular, shiny black female, is a puny runt by any standard, usually marked with distinctive grey and white stripes. Big dark knobs at the front of his head are the palps, his external genital organs. When he reaches maturity and the shiny black abdomen of a receptive female catches his eye—or his eight eyes—he climbs onto her web and vibrates his abdomen rapidly. On meeting, he spins a thin veil over her. It is tempting to think that this is an evolutionary survival trait, both of them knowing that she must be shackled to prevent her from turning on the weak and timid male before consummation of the sex act. But it probably has no such function and is purely a mating ritual, because she easily frees herself as soon as she is gratified. Meanwhile, the male plays out his unconventional role. First he spins a tiny web and deposits his seminal fluid on it, then proceeds to collect it on his knoblike palps and transfer it to the female's genital pore. The female is apt to immediately turn on him. If he is quick, he can make a getaway. Otherwise, he will provide nourishment for the eggs and ultimately for his offspring.

Once you see a black widow you are not likely to mistake her identity. Her abdomen, which is about 80% of her body, is globular, about the size of a large pea, and glossy black, which accounts for her having been called the "shoebutton spider." A prominent crimson marking shaped like an hourglass on the under side of her abdomen serves as a warning flag—the dash of sportsmanship frequently seen in nature. She inhabits dark or protected locations under stones, in brush, long grass, hollow stumps, or vacant rodent burrows. Unfortunately for human victims, she readily takes up residence in man-made structures such as cellars, garages, storage sheds, and outdoor privys. She is not happy in cold climates but, even there, succeeds in adapting. She is known in every state in the United States, as well as in Canada, and ranges south through Mexico and Central America to Tierra del Fuego at the southernmost tip of South America.

Wherever the black widow sets up housekeeping, she spins a coarse web and retires to a dark corner to wait for the vibrations of web that tell her a victim is enmeshed in the snare and is struggling to free itself. She quickly goes into action. She dashes out from her hiding place, but stops just short of reaching the victim, turns around and approaches her victim backwards. She exudes a strand of sticky silk from the glands at her rear end, extends the silk with her hind legs, and proceeds to hogtie the thrashing victim. If the victim is too violent, she ejects a large drop of sticky, viscous fluid that looks and acts like rubber cement. This quickly sets and subdues the most obstreperous prey. When this is done, and sometimes before, she administers the lethal venom, which may cause the victim to go into a frenzy and struggle violently to free itself. But the victim quickly weakens, gives up the death struggle, and dies. After sucking the fluids from the body of the prey, she cuts it loose and lets the carcass drop, usually leaving no trace of the struggle. She lives mainly on a diet of insects, other spiders, or any small arthropod that comes along. A single spider will devour an enormous number of small animals during her lifetime. Records kept by one student of spider lore showed the food captured by one spider totalled 250 houseflies, 33 vinegar flies, two crickets, and one small specimen of black widow.

When the black widow is ready to lay her eggs, she spins an inverted silken cup and fills it with eggs in a gelatinous film. She covers the open end with silk, forming an egg sack about the size of a small grape, spun of pure white tightly woven silk. When she is through, she has as many as several hundred small, pearly white eggs completely enclosed in a watertight bag. The spiderlings seem to hatch all at once, but that is because they spend several days in the egg sack before making their escape. From the beginning, they are full of vim and vigor, spinning little threads from pillar to post, dropping a line to a likely landing spot, and scampering back hand over hand for another try.

The young spiders grow steadily, and literally, by jerks. Spiders, like all arthropods, have a stiff external skeleton instead of a flexible skin. The young spider grows until its outer covering is stretched to capacity; then when its integument can stretch no longer without bursting, it splits and off it comes, and then the process of growth starts all over again. If food is abundant, the spiders grow amazingly fast. But if no other food is handy, they eat their little brothers and sisters with relish. On several occasions, I have collected black widow egg masses, placed them in jars to observe the hatching and development of the young spiders when fed small insects like houseflies. In one instance, through neglect, they received no food whatsoever. Amazingly, several of the young spiders from a brood of several dozen developed normally, but the brood became progressively smaller until only one fully grown spider was left.

The female black widow is notorious for inflicting painful, sometimes debilitating, bites to incautious people, especially in rural areas where the spiders tend to lurk in dark places of outhouses. Men's genitals are attractive targets. Men in that situation, or women for that matter, might well heed Dicken's curt warning in his *Pickwick Papers*, "Beware of widders."

Insects are the most numerous terrestrial animals on earth; they have the largest number of species and are the most successful with respect to the variety of habitats they occupy, which is almost every conceivable niche on earth. Their beneficial and harmful influence on human life is enormous. The fossil record shows insects appearing on the scene more than 300 million years ago and thriving through the time of the massive extinctions of the dinosaurs and other animals during the Cretaceous, roughly 100 million years ago. The insects appeared hundreds of million years before the first primates, and in view of their amazing versatility and adaptability, there is reason to believe that they will still be thriving after the human primate becomes extinct. Today, nearly a million species of insects make up more than three fourths of all animal life on earth. It is not surprising that insects have acquired through evolution an enormous variety of forms, habits, and life styles.

The typical adult insect has six legs, two pairs of wings, and three main body parts: head, thorax, and abdomen. But there are pronounced differences between groups. Some have only two wings, and some none at all; the larvae of some species are legless, and others have many legs. Some species have returned to life in water, and in others only the larvae are aquatic. Many have become parasites or predators of other organisms, their victims ranging from other insects, to plants, humans, and other animals.

Aphids, so-called plant lice, are among the most prolific of animals, and part of their success is due to the fact that they regularly switch from sexual to asexual reproduction when there is an advantage to it. Summer generations are exclusively females that are parthenogenetic (without male fertilization), usually born already hatched within the female (ovoviparous). Typically, males appear in the fall and mate with females that lay fertilized overwintering eggs. But in some species no males at all have ever been seen. Parthenogenesis gives the organisms a tremendous advantage in conservation of energy and in minimizing the chance that mates might not find each other. Animals such as aphids that reproduce parthenogenetically can be seen to increase in numbers very quickly, to the dismay of farmers and home gardeners. But the long-term outlook for the survival of species that reject the benefits of genetic diversity derived from sexual reproduction is questionable at best. The life of an aphid that lives on grapevines is one such puzzle.

Most wine lovers have had the pleasures of the palate affected one way or another by the sex life of the phylloxera, *Phylloxera vitifoliae*, because of its destructive effect on some varieties of grapes. The phylloxera is a native of the eastern United States where it feeds on wild and cultivated grapevines. It is hardly ever a problem there, although it commonly causes galls on the leaves. But when immigrant French and Italian vintners brought European varieties of grapes to the western part of the country, phylloxera attacked the vines with a vengeance by going underground and causing knots and galls on the roots, weakening the vines and often killing them in 3 to 10 years. When the aphid arrived in France in the 1850s on infested nursery stock, it caused a devastating epidemic, wiping out more than 2.5 million acres of vineyards. The remedy against the phylloxera was to replant vineyards with vines grown on resistant rootstock native to the eastern United States. Improved rootstocks were developed in California, but the problem did not go away because either the selected stock was less resistant than at first thought, or more probably, the aphids evolved to adapt to them. As this is written, many vineyards are being replanted.

The phylloxera has a complicated asexual and sexual life cycle, with four distinct forms of adults. In the eastern United States, the aphids spend the winter both on the roots and as eggs on the aerial canes of grapes. The eggs hatch in the spring, and the young feed on the tender leaves where they form galls and give birth asexually to live young for several generations. Some of them drop to the ground and feed on the roots for several more generations. In the fall, winged females appear on the roots. They work their way to the surface, fly to the upper parts of the vine, and lay eggs that hatch into sexual males and females. After mating, each female lays a single egg that remains on the vine through the winter.

The evolutionary advantage of the sexual stage in the phylloxera's life cycle is not immediately apparent. The parsimony of laying a single egg, the product of the efforts of a male and a female and generations of their parthenogenetic ancestors, reflects an extreme case of conservation of energy, evidently for an important gain. They do not need fertilized eggs to survive the winter because aphids remaining on the roots also live through the winter. Possibly they need at some stage the kind of nutrients they get from the leaves to sustain the life cycle, and the egg stage is merely incidental. Is there something more compelling—an imperative need to rejuvenate the long line of clones with a reshuffling of genes? Students of the aphid noticed that there was a progressive decrease in fecundity in succeeding asexual generations of the aphids until the final, sexual form was produced. One of the investigators, M. Balbiani, who discovered the sexual eggs more than a century ago, proposed a theory that if it were not for the renewed vigor resulting from the fertilized egg of the sexual form, the phylloxera would soon become extinct. The idea is a persistent theme among biologists. To paraphrase Harvard biologist Ernst Mayr, "Any organism that becomes asexual is apt to become extinct sooner or later." The other side of the coin in the case of the phylloxera is that in California the sexual eggs appear, but the aphids hatching from them do not seem to mature, whereas the asexual forms seem to reproduce indefinitely. They have evolved to live from year to year only on the roots. After a history of more than 150 years of study, the jury is still out on the significance of the sexual form of the Phylloxera and what its effect, if any, is destined to be on the long-term fate of the species.

Many insects are intimately tied in with their hosts, and their reproductive lives are so rigidly programmed that a long association is evident. That kind of adaptive situation has been called *coevolution*, relating to the fact that two species, such as a parasite and its victim, can evolve in a joint relationship that is essential for survival of at least one of the pair. Paul Ehrlich made a study of the coevolution of butterflies and plants, and his findings on the populations of species may have stimulated his well-known interest in human populations. Coevolution is seen in orchids, some of which have flower parts that mimic female bees. When male bees visit the flowers for food, they try to copulate with the flower parts that they think are female bees. Their movements are effective in transmitting the orchid's pollen.

Fig wasps, *Blastophaga psenes*, have a sexual relationship that must have evolved from a long history of coevolution. The tiny, torpid male is born, lives, loves, and dies in the confines of the interior of a fig. The active female is a small, shiny black wasp, no longer than 2.5 mm (about 0.1 inch) and, as typical of bees and wasps, with four wings on which she wings her way to the outer world. The male is a pale, larvalike, wingless insect with a long, slender abdomen. He never sees the light of open day with his multiple, but poorly developed eyes, but instead waits in the dark for the female, whose purpose in life is to parasitize fig seeds. She inadvertently performs a service to mankind by helping horticulturists produce tastier figs. This is related to the peculiar structure of the fig and the fig's dependence on its own friendly enemy.

A fig is a hollow globelike receptacle with an inner lining packed with minute flowers pointed inward. The flowers must be fertilized to produce seeds, important also from the human viewpoint because without seeds the fig does not develop its sweet, nutty flavor. But the fig has a problem. The flowers cannot self-pollinate, and the only entrance to the flowers is a very small opening at the end of the fruit, an unlikely prospect for pollen to reach the hundreds of enclosed flowers by floating on air currents. When a female wasp finds her way into a fig, she deposits an egg in the ovary of each flower she comes to for as long as her supply of eggs lasts. She may have as many as 300, but this is only a fraction of the number of flowers inside the fig. The egg hatches, a larva grows and develops within the ovary, forming a *gall* where a seed would ordinarily form.

The male matures first and works his way out before his female mates emerge. When he finds a gall containing a female, he gnaws a hole through the thin, translucent membrane at the top of the ovary and gains access to the female while she is still inside. Her tiny cell is too crowded for him to enter, so he mates with her by inserting his long, slender abdomen through the opening. After impregnation, she pushes her way out through the hole made by the male and finds her way to the outside of the fig. Fortunately, the males either cannot find all the females, or they grow weary of the quest before all the females are fertilized. As with all other wasps and bees, fertilized eggs develop into females, and unfertilized eggs develop into males. The sex ratio of fig wasps is about 16 males to every 100 females.

The natural host of the fig wasp is the wild fig known as the *caprifig*. During the female's escape to the outside, she becomes covered with pollen from stamens of the flowers in the caprifig. After gaining freedom she flies about looking for a newly developing fig and works her way inside through the small opening at the end. Inside, she crawls around, laying eggs in the single ovaries of the flowers by inserting her ovipositor into the pistils, all the while pollinating flowers as she crawls from one to the other. She can pollinate the flowers of only one fig because her wings are scraped off when she squeezes her way through the tiny hole at the eye of the young fruit, and she will die there. The caprifig could not survive as a species without the fig wasp. Self-pollination is impossible because the staminate (male) flowers do not produce mature pollen until several weeks after the pistillate flowers in the same fig are receptive. But figs in different stages of development make it possible for the wasp to cross-pollinate.

It is here that humans enter the equation. The Smyrna fig is a highly prized variety of superior eating quality, but it produces only female flowers and therefore cannot be fertilized unless the inferior caprifigs are growing nearby, a fact well known to the ancient Greeks. About 340 BC Aristotle described a small creature called *psen* that entered unripe figs and caused them to remain on the tree until maturity. Aristotle's student, Theophrastus, described the method, although he did not understand how the wasp caused the seeds to develop. He thought the fig was a flowerless plant and that the wasps made cultivated figs swell and grow to maturity by nibbling on them.

What was well known to the ancients was greeted with skepticism and laughter when attempts were made to grow superior varieties of figs in the United States. Even scientists in Italy ridiculed the idea. But when the truth was finally accepted, it became possible to grow figs equal in quality to the Smyrna figs previously available only from the Middle East. The procedure is called *caprification*, most efficiently done by placing baskets of caprifigs in the branches of Smyrna fig trees. If caprifigs are nearby, some of the pollen-bearing wasps that emerge from them will find their way into the Smyrna figs and pollinate them. Several hundred females may emerge from a single caprifig, and because a mature tree may bear a spring crop of as many as 20,000 figs, enormous numbers of female wasps are available for caprification.

Caprification depends on a case of mistaken identity on the part of the female wasp. She cannot lay her egg in the ovary of the Smyrna fig because the Smyrna has a long-styled flower, making the distance to the ovary too far for her to reach it with her ovipositor. She dies in an abortive attempt to perpetuate her lineage. But meanwhile, pollination has been accomplished. Thus the sex lives of figs and wasps and the appetites of humans are intertwined in mutual dependence—to the delight of lovers of Fig Newtons.

Historically, honey bees, *Apis mellifera*, have been a favorite symbol of fertility, industry, and zeal. In Salt Lake City there is a large house with a structure on the roof shaped like a beehive. The house was once the residence of Brigham Young, founder of the Mormon settlement in Utah. The building is known as The Beehive House. And a beehive forms the motif of the Great Seal of the State of Utah. A similar design was used by the United States government when it issued a $45 bill on January 14, 1779, on which there was depicted an apiary with two beehives and bees swarming about. The motto inscribed on the bill was *Sic Floret Republica*, meaning *Thus Flourishes the Republic*.

The social insects are the most sexist societies on earth. The worker honey bees that can be seen buzzing from flower to flower collecting pollen and nectar are females deprived of sexuality. They are one of three castes in the bee society that can have as many as 50,000 individuals in a single hive. The other castes are the drones (males) and the queens, and these are few in number. The bee's sex life is more complex than the "birds and the bees" myth implies. There are almost always more of the sex types—drones and queens—than needed to perpetuate and maintain the stability of the society. When several young queens mature at the same time, it at once becomes apparent that there can be only one queen. They attack each other in a pitched battle for supremacy, and invariably it is a fight to the finish with death to the vanquished. The last survivor wins the right to be the first in line to wear the crown as head of the colony. John Henry Comstock, a distinguished entomologist, told of one morning when he found the lifeless bodies of 15 murdered queens that had been cast forth from the hive.

The surviving queen mates in flight, usually about the sixth day after she reaches maturity, and on her first or second flight from the hive. During the so-called nuptial flight, she may copulate with one or several males. But even if she mates only once, a single drone can ejaculate 5 or 6 million sperm, enough to last for the rest of her life. She stores the sperm in her body for use throughout the 4 or 5 years of her life to fertilize the eggs—except those destined to become drones—as they are formed.

The drones are larger than the workers and not at all popular with the busy sexless females. The drones are loafers, useless as workers, and do nothing but eat. There are almost always more drones hanging around than are needed to mate with the occasional queen that requires their services. They run the risk of being attacked by vengeful workers in late fall or early winter or at times when food is scarce. After a drone's nuptial flight with a new queen, when the two separate, the tight grip of the queen's claspers tears out the drone's genitals, allowing time for the sperm to transfer even after the drone's death. After mating, the queen settles down in a secluded part of the colony and for the rest of her life does nothing but eat food that the workers bring to her and squirt out eggs. She has the authority and the biological equipment to determine the number of potential drones in the hive. The workers of the Hymenoptera, as bees and ants are called, all come from fertilized eggs, and there are plenty of sperm to go around. The queen has the ability to withhold sperm from an egg, in which case the unfertilized egg develops into a male. The result is a complicated ancestry. A drone has no father, but he has a maternal grandfather who, in turn, was fatherless, and so on.

The queen, pampered by her attendants, becomes an egg factory.

She starts laying eggs 2 days after mating and continues to spew them out in prodigious numbers. She lays several times her own weight in eggs every day for weeks on end. She may lay as many as 2000 to 3000 eggs in a day, or more than a million during her lifetime. Still, she is no match for a termite queen, who can turn out more than 80,000 eggs a day until millions have been laid.

The queen bee, like her sister workers, came from a fertilized egg, but she develops differently because she gets a different diet. She receives a food called *royal jelly*, which contains a hormone that promotes sexual development. None of those destined to become workers get any of the royal jelly except during the first 2 or 3 days of their larval lives. The workers get a low-protein diet that keeps them sexually immature, sterile, and willing workers. The queen gives off odors—chemical signals called *pheromones*—that block sexual maturity by inhibiting ovarian development in the females that are destined to become workers. The virgin queen also produces a perfume that gets quick action from drones. As many people know from painful experience, the worker bee has a stinger, part of her otherwise unused reproductive equipment. The queen also has a stinger, but she uses it only to kill rival queens.

Birds are among the most lovable—and loving—animals, a fact that belies their reptilian origin. The fossil record clearly shows that birds evolved from reptiles about 150 million years ago, probably from the bipedal Archosaurians. The bipedal feature left their forelegs free to develop into wings, but whether flight came about from flapping their "arms" to help them run faster or to glide from cliffs or trees is a matter of lively speculation. And how supposedly cold-blooded terrestrial reptiles became transformed into warm-blooded flying birds leaves questions that may never be completely answered. One view is that the birds' reptile ancestors were warm-blooded. An intermediate fossil form is the famous *Archaeopteryx* that had feathers but had reptilian teeth in both upper and lower jaws and a generalized lizardlike build. Its form suggests that the earliest birds lived in forests where they were capable of running, jumping, and gliding among the branches. Recent evidence indicates that the ancestors of birds were arboreal (tree-living) reptiles.

Internal fertilization of birds is an advance over the typical amphibian's dependence on external fertilization. But unlike their distant relatives—lizards, snakes, and other modern reptiles—male birds for the most part have no really good equipment for sexual intromission. Male and female birds are equipped with a cloaca, an ancient device seen in many primitive animals from worms to toads, consisting of a common opening for genital and excretory products. In birds, however, special features show how far removed they are from their ancestral aquatic environment. A system for concentrating the combined urinary and fecal material conserves water by reabsorbing it. This economy enables desert birds to live for long periods without water.

The *vas deferens* of the male bird ends up in an erectile *papilla*, the only copulatory organ of most male birds. During copulation, both male and female birds evert the lip of the cloaca called the proctodeum, which they press together, and the male squirts his sperm into the part of the female's urogenital tract called the *urodaeum*, from where the sperm find their way into the oviduct. A few species of birds have a male penis and a female clitoris, present in ducks and swans, and in the group of flightless birds, the so-called ratites, that include ostriches, the emu and cassowary of Australia, the kiwi of New Zealand, the rhea of South America, and the now extinct moa of New Zealand. There are more than 25,000 known species and subspecies of birds, with great diversity in form, color, and behavior. They display a remarkable variety of courtship rituals that in some species are elaborate and prolonged. Bird songs and their meaning are a delight and a study in themselves. The performance of many species of birds in protecting and caring for the young in terms of dedication and energy expended is a remarkable illustration of the evolution of sexual behavior in action. Emperor penguins are a grand example.

The courtship of emperor penguins, *Aptenodytes forsteri*, at their rookeries (nesting sites) in Antarctica is probably the coldest tryst on record. These dignified monarchs of the bird world make love, "nest," hatch, and rear their young where the climatic forces are such that "No words can express its horror," according to Antarctic naturalist, Cherry-Garrard, as quoted by Katherine Muzik and Michael Fedak, who studied the penguins' ordeal. In contrast to the frigid Antarctic environment of blizzards, hurricane-force winds, and an average minimum temperature as low as −48°C, the warmth of the emperor penguin couple's loyalty and devotion and their dedication to raising and protecting their chicks in an incredibly hostile environment during the coldest part of the year is a monumental case of reproductive determination and sacrifice.

Emperors are the largest of the penguins, standing 1.3 meters (4.27 feet) tall when walking with their necks extended, and weighing 20–40 kg (44–88 lbs). They are the only penguins that never go on land, but spend their entire lives on the ice and in the Antarctic sea. They are superb swimmers, using their modified wings, or flippers, to dive beneath the surface of the sea where they feed on such seafood as fish, squid, and a small, shrimplike animal called krill. On the ice, they rest, preen, mate, hatch their eggs, and care for their young. They devote the Antarctic summer to feeding on the abundant supply of sea food and putting on fat in preparation for the arduous winter ahead. In March, which is the onset of the Antarctic autumn, they leave the open water and start across the ice on foot to their rookery. The emperors have a scattering of about 30 rookeries around the periphery of the Antarctic continent. They may have to walk, single file, as far as 120 km (75 miles), depending on how far the rookery is from water.

One of the emperor's rookeries is at Cape Crozier on Ross Island, at the edge of the Ross Ice Shelf. It is not far from McMurdo, the site of the large research station at the southern tip of the island. It is a long walk for penguins on their stubby legs. Although penguins are skillful swimmers, they are awkward out of water, and make headway slowly on ice. But despite their ungainly gait, they are able to walk several miles a day at an average speed considerably below their top speed of 2.8 km (1.74 miles) per hour. When they want to go faster they are able to do so for short distances by skidding on the ice. They belly-flop and act like a toboggan, using their feet and flippers to propel themselves along. In that way, they can go about as fast as a human can run on the ice.

They court and mate in April. By May, the female lays a single egg that weighs about 450 grams (about 1 lb) and at once returns to the sea to feed, leaving her mate to keep house in the coldest place on earth. He incubates the egg by holding it balanced on his two feet and covered with a fold of skin on his bottom. He holds the egg in this position as long as 65 days. Groups of males with their eggs huddle together, as many as 5000 to 6000 of them in a group at some rookeries, for what protection they can give to each other against the bitter wind. During the huddle, there will be about 10 penguins per square meter. Huddling cuts the drain on their reserve fat supply almost in half and doubles the bird's endurance. In July—the middle of the Antarctic winter—about the time the egg is ready to hatch, mother penguin returns with her belly full of seafood and finds her mate by calling. He recognizes her call and returns it. She takes the chick from its father, holds it on her own feet, and regurgitates the undigested food in her stomach to give the hatchling its first meal.

By now, the male has been without food for nearly 4 months, for the last 2 months or more sitting stoically hunched against the fury of the Antarctic winter. After his long fast, his reserve supply of fat is nearly exhausted, and he may be down to nearly half his normal weight. At last, he is free to go back to the open sea for that long-awaited fish dinner. After 4 weeks, he returns, and thereafter, the couple take turns keeping the chick warm and feeding it while maintaining a shuttle service between the rookery and their source of food in the open sea. When the ice begins to break up under the warming sun, the youngster, now called a fledgling after about 5 months of tender loving care, is ready to leave the nest to make it on its own.

The winter courtship and family life of the emperor penguin involve a terrible risk for which there must be an evolutionary benefit in return. When the emperors start their walk to the rookery, they have a good store of fat in reserve. But if during their ordeal they use up the reserve, they will have to abandon the egg or young and return to the open water for food, in which case the egg or young will quickly freeze to death. Tests show that if the male's body weight goes down to 22 kg (about 48 lb) before the female returns, he will abandon the egg or chick and go to the open sea to feed. When that happens, the penguins would have wasted a year's reproductive effort, and the population would be in jeopardy of losing out in the constant struggle for survival in one of the most unfriendly environments on earth. The chicks suffer a frightful mortality. A loss of 90% was recorded at one rookery during the winter of 1972.

Why do they run this risk? Why do they walk a distance of 200 kilometers (124 miles) or more, and go without food for 100 or more consecutive days? Why do they nest at a time that requires the eggs to be incubated and the chicks to be nursed in the dead of winter? Why do they go through the phases of the reproductive cycle at a time during the Antarctic winter when they must huddle together to survive? Why not breed and rear their young in the spring and summer like most sensible animals?

One proposed answer is that hardly any predator would venture out in the Antarctic winter when the emperors are nesting. Eggs and chicks would be especially vulnerable to predators when the large numbers of gregarious birds are nesting close to one another. But the emperors are not known to have any land predators, although skuas and giant petrels might be hazards during the warmer part of the year. A more important reason might be that by brooding in the winter, the chicks are ready in the spring to brave the waning rigors of the ice and go out on their own when air and water are warmer and the Antarctic ocean is teeming with seafood.

When fishlike amphibious creatures abandoned the sea for the unknown hazards of terrestrial life, they had to overcome innumerable problems, one of which was how to make sperm and eggs come together. They no longer had the convenience of an expansive swimming pool in which sperm could swim around looking for or bumping into eggs. A simple solution would have been to leave sex behind with the other rejected habits of their past life, and become asexual. Perhaps some of them did, but if so, few of them survived. Instead, nearly all of them were compelled by the sex imperative imposed by the forces of natural selection to devise ways of reproducing sexually in a dry environment. Different ingenious and innovative methods were invented by different species. Spiders, insects, and birds found ways of reproducing that suited their anatomy and way of life. Some of them, like the emperor penguin, expend enormous energy and suffer agonizing loss of life, yet submit under the most extreme conditions on earth to the evolutionary forces that compel them to comply with the sex imperative.

Chapter 7

The Penis Innovation

The penis is the quintessential copulating organ of male terrestrial animals. The device was invented independently in such diverse forms as the simple intromittent organ of planarian worms, the two-way system of snails, the stringlike penis tube of barnacles, the individualized organs of insects, the rare penile extrusion of frogs, the double penis of lizards, and the few innovative birds that managed to partially transcend the crude expedient of cloacal kissing. The birds' tentative effort to do better was confined mainly to birds that evolved from terrestrial reptiles to fliers and back to the ground again with useless wings.

Copulation is not ordinarily thought of as' being part of the reproductive pattern of fishes, in which the usual method of spawning is for the female to extrude her eggs in a suitable location where the male follows and emits his milt in the vicinity of the eggs. But there are many variations. Some fishes manage to fertilize the eggs internally; one of them is the common guppy of the home aquarium in which the male has a fin modified into an intromittent organ. The fertilized eggs remain in the female until they hatch, so the young guppies are "born alive." (Quotes are used around "born alive" because the alternative would be "born dead.") Animals in which the eggs hatch within the mother are called *ovoviviparous*.

Arthropods—animals with jointed appendages—are the dominant animals on earth, outnumbering all other groups combined. They include such animals as lobsters, crabs, centipedes, spiders, and mites. Arthropods probably evolved from the same ancestors as the soft-bodied marine segmented worms. The arthropods' jointed exterior skeleton, called an *exoskeleton*, requires specialization of their copulatory organs to fit each species' rigid structural configuration. The sexual organs are almost always distinctive for a species and, in many cases, are useful for identification. One view is that a slight modification in the lock-and-key arrangement of the male and female genitalia is enough to create a separate species without any apparent difference in superficial structures. Two species of tiny spider mites that attack trees, shrubs, and small plants—the two-spotted mite, *Tetranychus telarius*, and the Pacific mite, *Tetranychus pacificus*— look so much alike that one can rarely tell them apart even with a magnifying glass. The two species are about the same size, have similar habits, and in many cases attack the same varieties of plants. The only way to tell them apart with certainty is for an expert to examine the male genitalia under a microscope. Mite and insect penis experts call it an *aedeagus*, from the Greek *aidoia*, meaning *genitals*.

The arrangement of the genitalia of some arthropods is highly innovative. Robert Husband of Adrian College described a species of mite in the family Podapolipodidae that infests beetles. It lives between the folded outer wings (called elytra) and the back of the beetle's body. The mites' genital organs are shifted upward to their backs so the male and female can copulate back-to-back while positioned on opposing surfaces. A similar adaptation is seen in another species of mite, *Rhynchopolipus rhynchophori*, that infests palm weevils. The aedeagus projects like a long spear over the male's back, ending in an arrowhead point in front of the mite's head.

Male bedbugs have a foolproof way of getting their sperm to the target while, at the same time, minimizing the chance of their effort being nullified by subsequent males. The male of the common human bedbug, *Cimex lectularius*, pierces the female's integument with his spearlike penis and puts a hypodermic injection of his sperm into her body cavity. The sperm migrate through the haemolymph—the insect's blood—to the ovaries. Females of some species that receive

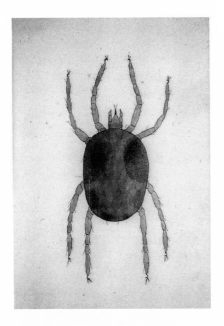

Some species of spider mites are so similar that the only way to identify them with certainty is to examine the aedeagus (penis) with a microscope.

hypodermic insemination have special reproductive structures to receive and transport the sperm.

A penis more closely identified with the male copulating organ in higher vertebrates was invented by ancestors of modern lizards, snakes, and related reptiles. Exactly when the vertebrate penis appeared on the evolutionary scale is a mystery. Birds, who share an early origin with reptiles, branched off from a remote ancestor before the penis became a reptilian feature. There is no way of knowing how the giant dinosaurs like *Brontosaurus* managed to copulate. If their method was birdlike, it is conceivable that ponderous behemoths had problems maneuvering into position. The only living reptile that does

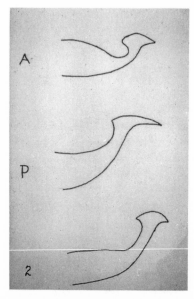

Penis of the two-spotted mite, compared with penises of the Pacific mite and Atlantic mite.

not have a penis is the so-called living fossil, the sphenodon or tuatara, a large lizard that inhabits islands off New Zealand. It grows to
a length of 2 feet and has hardly changed at all in the last 140 million
years.

It is possible, but improbable, that some or all of the dinosaurs
were parthenogenetic. If so, the lack of sexual reproduction for
generations on end over millions of years would conceivably contribute to their demise. Present-day gecko lizards, *Hemidactylus garnotti*
and *Lepidodactylus lugubris*, are asexual, reproducing without benefit of males. They were largely replaced by a related sexual lizard, the
house gecko, *Hemidactylus frenatus*, when it was introduced to
islands of the tropical Pacific Ocean. But the decline of the native

geckos is probably due to the small natives avoiding the larger invaders. Curiously, many geckos are loudly vocal, making clicking and cheeping sounds. They are the only members of the Squamata (lizards and snakes) that lay hard-shell eggs. Those of all the others have a leathery texture.

An evolutionary step in development of the penis is seen in the anole lizard, *Anolis carolinensis*, which is doubly endowed with what appears to be nearly a foolproof way of fertilizing the female. The anole is commonly called the American chameleon; it is not a chameleon at all, but a small iguanid lizard. It makes its home throughout the southeastern United States where it inhabits moist hardwood hammocks, palmetto, and similar scrub. Anole lizards have developed a courtship behavior that influences the reproductive physiology and behavior of the opposite sex, similar in many ways to what we find in higher animals. As long as males are still fighting over territory in the female's presence, her ovaries do not grow, but if exposed to frequent courtship, she has a high rate of ovarian development.

When the male confronts a female that he thinks is interested, he goes through a repertoire of up-and-down bobbing movements coordinated with the extension of a bright red dewlap, a fold of skin (called a gular fin) that hangs under the chin. Mating follows a characteristic pattern. A sexually interested male approaches a female and goes through his courtship sequence. If the female is ready she arches her neck and allows the male to grasp it. He throws his tail under the female's tail and presses his genital opening against hers. He is now in position to penetrate. The male has a double penis, consisting of two parallel penile organs called hemipenes, but he uses only one at a time. If he swings his tail to the right, he inserts his right hemipene, and if he swings his tail to the left, he uses his left hemipene.

Harvard biologist David Crews found that most males of the lizard, which he calls the annotated anole, do not have a preference for either the right or left hemipene. They are sexually ambidextrous, using whichever hemipene is handier, but a few males have a preference, being either right- or left-pened. When Crews surgically removed the preferred hemipene, the hemipenectomized lizards readily switched to the other pene without a hitch.

The annotated anoles normally copulate in some exposed place, such as a tree trunk, fence post, or the side of a house. Copulation is usually prolonged, aided by the barbed configuration of the penis, and this makes the lizards easy prey for birds and other predators. A copulating pair can be picked up easily by hand without separating. It is a risk the male is instinctively compelled to take, because the time lapse gives him an advantage in the one-upmanship game of sperm competition in case other eager male lizards are lurking in the vicinity.

The hemipenes of lizards and snakes are two parts of a bilateral reproductive system. The seminal fluid does not empty into a common receptacle as in other terrestrial vertebrates. Each hemipene has its own vas deferens leading from its own testis. Nor is the sperm ejected through a urethral tube extending through the length of the penis as it is in mammals. The seminal fluid is delivered from the vas deferens into a groove on the surface of the hemipenes. Crocodiles and tortoises have a single penis, but delivery of the seminal fluid is by a similar means—in a groove that serves as a conduit running the length of the penis. The hemipenes of lizards are apt to be spectacularly ornamented with nipples, whorls, and spikes, the shape and arrangement of which are sometimes used by systematists to classify species. Erection in lizards and snakes is by both vascular engorgement and muscular action. In some snakes, the sperm can be held by the female for many months, as evidenced by female snakes laying fertile eggs several months or even years after being captured and isolated. As in higher vertebrates, the female's ovaries alternate in producing a single egg, in the case of the anole, every 10–14 days (20–28 days for each ovary).

The prolonged coitus of the annotated anole is not unusual. In contrast to the brief encounter between males and females of some species, there are many in which the couple remains connected for several hours. Some insects are held together mechanically by highly complex anatomical structures on their abdomens, and they remain locked in their position while the sperm is being transferred. Males of the butterfly *Parnassius* and a few other species produce a secretion that hardens to a cement and holds the mating couple together during the period of sperm transfer. The binding can be so strong that physically uncoupling them tears out the male genitals and is fatal to the male. In some species, the hardened secretion remains in the female, serving as a barrier that prevents other males from copulating with her.

A case in which the insertion of a sperm plug is only "wishful thinking" on the part of the male is that of the damselfly, a smaller and more fragile relative of the sturdy dragonfly. These insects were abundant 300 million years ago, before the big dinosaurs arrived, and have changed very little since then. The copulating organ, or so-called pene, of the male is on the second segment of the abdomen near the base instead of at the tip and is separate from the vas deferens. The penes are species specific, but they all have hooklike projections of various shapes that are used to scoop out any sperm that has been deposited by a previous male.

In canines, the base of the penis swells up after ejaculation, and the female's vaginal opening constricts, making it impossible to withdraw for 10–30 minutes or longer. A bitch in heat will attract all males within sniffing distance, which is considerable, and will give them a merry runaround until one of them is permitted to mount her. The competitive strategy of getting stuck allows time for the sperm to reach a favorable location to fertilize the ovum before any of the other males can enter her. The adaptations that prevent or delay impregnation are seen by biologists as strategies to eliminate sperm competition and protect the successful male's investment. Typically, males take every opportunity to utilize their full reproductive potential. But unless the progeny or other members of the family are dependent on him for food and protection, nature has little concern for the welfare of the father after he performs his reproductive task.

In mammals, there is great variation in the size, shape, color, and location of the penis, but they all, whether of mice or men, have one thing in common. The mammalian penis is designed for both urinating and copulating. The failure of evolution to follow through on a complete separation of the primitive cloaca's three-way function—anal, urinary, and genital—can cause the male a great deal of distress. As aging progresses, most males suffer an enlargement of the prostate—a gland that surrounds the urethra—which, whether in horses or humans, is apt to restrict the flow of urine to the point of being dangerous to health.

The penises of mammals have evolved into three different modifications. The basic design in many mammals, including rodents, canines, horses, elephants, and humans, is an appendage containing vascular tissue that becomes engorged with blood under the stimulus of mental or physical sexual excitement, causing the penis to become longer, larger, and hard. The arteries are dilated allowing blood pressure to fill the spongy tissue; the veins are simultaneously constricted to impede the flow of blood away. A constrictor muscle further retards the escape of blood, having the overall effect of an erection.

Both nerve action and hormones are required to bring about an erection. Remarkably, a simple inorganic chemical, nitric oxide, a toxic substance found also in other parts of the body in minute amounts, is required for an erection. Its action is that of a neurotransmitter that makes it possible for a nerve impulse to travel from one nerve ending to another. An erection is quicker or slower depending on a variety of psychological and physiological circumstances.

The penis in some animals, including cattle, sheep, goats, and deer, is quite different from that previously described. In these animals, the penis is a sheathed, elastic, gristly shaft that is not primarily dependent on blood pressure but is unsheathed by muscular action and is instantly ready. These animals are usually able to consummate impregnation quickly, often within a few seconds. During the days when most heavy work was done by horses, oxen, and, unfortunately, slaves, instead of cars, trucks, and tractors, it was a practice to surgically extract the penis, called the pizzle, from butchered bulls for use in punishment or unfriendly persuasion, hence the name "bullwhip."

The third evolutionary advance in penile structure was the addition of a penis bone or *baculum*, a feature in rodents, bats, seals,

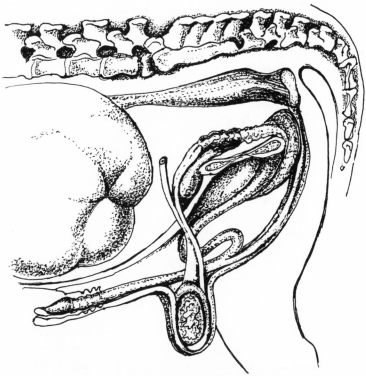

Reproductive system of the bull, typical of a group of animals including cattle, sheep, goats, and deer. The penis does not contain vascular tissue; instead, it is a long, gristley, cartilaginous rod held in a sheath, instantly ready by muscular extrusion. Redrawn by Vicki Frazior from Dadus M. Hammond, Paul R. Fitzgerald, and J. LeGrande Shupe, "Trichomoniasis of the Reproductive Tract," in *Animal Diseases, The Yearbook of Agriculture 1956*, U.S. Department of Agriculture, Washington, D.C.

sea lions, walruses, other carnivores, and some primates. The baculum lies on top of the vascular tissue, helps to keep the penis rigid, and takes over a good part of the action before tumescence (swelling with blood) comes into play. The penis bone is very large in pinnepeds (seals, sea lions, and walruses). It is especially large in walruses. Most of the species of seals copulate in the water, in which environment the penis bone is especially helpful. Why the genetic shuffle did not provide humans with this handy device is not known. In view of the common occurrence of male impotence and reported decline in penile rigidity during attempted erection with advancing age, a baculum would seem to have considerable utility. On the other hand, from an evolutionary strategy view, the species has a better chance of surviving and thriving if a woman is impregnated by a young, virile male, who is less apt to produce sperm with deleterious mutations or other abnormalities. In view of the vaunted and some- times flaunted sexual supremacy of the human species, the presence or absence of a baculum may be of trivial evolutionary significance. But if the reports (by clinics engaged in treating male malfunctions) that 10% of men are impotent are accurate, where the baculum gene or genes went would be of interest to millions of people.

Much has been made of the differences in size of the external genitalia of various animals, especially primates. Robert Martin of University College, London, and Robert May of Princeton University pointed out that there is a relationship between the size of the male genitals and dimorphism (differences between males and females in physical attributes, especially size), which in turn is related to whether the animals are monogamous or polygamous. The male gorilla, *Gorilla gorilla*, is much larger than the female. He will most likely have three or four females in his harem, each of which will have about a 4-year interval between births. The male's penis when flaccid is invisible, being hidden by the prepuce, forming a depression in the skin. Even when erect, his penis is only 3 cm ($1\frac{3}{16}$ inch) long, and his testicles are the smallest of any of the higher primates. The male orangutan, *Pongo pygmaeus*, is about twice the size of the female. She comes in heat only once or twice in 6 years, giving birth only every 5–7 years. This means that copulation comes close to being nonexistent, although the mating system is essentially that of a small harem. The orangutan's erect penis is about 4 cm ($1\frac{1}{2}$ inch) long.

The sex life and the sexual apparatus of the chimpanzee, *Pan troglodytes*, are in sharp contrast to those of the gorilla and orangutan. Female chimps are notoriously promiscuous, and there is little sexual dimorphism. The males have the longest penis of the great apes, but still not as long as that of humans. The chimpanzee's erect penis is about 8 mm ($3\frac{1}{8}$ inches) long. Its bright pink color serves as a copulatory display, revealing the male's intentions to the female, for he always approaches the receptive female with an erect penis and copulates as soon as they make contact. There is moderate sexual dimorphism in humans, the male being on average about 20% larger than the female. But alone among the higher primates, humans have the distinction of engaging in all the mating systems: monogamy, polygamy, and promiscuity, although monogamy is held to be the ideal in most societies. Several primatologists have pointed out that the human male has the largest penis of all the higher primates. R. V. Short of the University of Edinburgh, citing W. A. Schonfeld, who made an extensive survey, states that the human adult penis averages 13 cm ($5\frac{1}{8}$ inches) in length. Measurements were made on the fully extended flaccid penis, presumed to be equal to its length when erect. The longest 10% in the survey averaged about 16 cm ($6\frac{1}{4}$ inches).

A. H. Schultz showed in 1938 that there are great differences in the size of the testes of the great apes in proportion to their body size, and R. V. Short determined that the differences were dependent on their breeding habits. Female gorillas and orangutans in the wild are thought to mate normally with only one male, so they copulate infrequently. But when female chimpanzees are in heat, they typically mate repeatedly, and with several males. In chimpanzees, selection should favor the male who can produce the most sperm, so theoretically, chimps should have larger testes per body weight than do primates with single-male mating systems. A. H. Harcourt of the

University of Cambridge and co-workers compared the testes weight of a large number of primates and found the hypothesis to be correct. For example, the ratio of testes weight to body weight of the chimp is 13 times that of the gorilla, and $4\frac{1}{2}$ times that of humans, but even so, humans have a ratio of testes to body weight 3 times that of the meagerly endowed gorilla.

There is also a difference in sperm production depending on the size of the testes, and therefore, on the volume of seminiferous tubules. The multimate rhesus macaque, *Macaca mulatta*, produces 23 million sperm per gram of testis per day, compared to humans, who produce an average of only 4.4 million sperm per gram per day. Among the higher primates, chimps are the champs, putting out more than twice the number of sperm per ejaculate than the human male.

There is a mystique about the size of the human male penis that prompted a number of studies, most of them of questionable reliability. The emphasis is almost always on the length of the appendage. An assumption is that there is either an evolutionary advantage to ejaculating the seminal fluid deep in the womb or that the greater the length, the greater the woman's pleasure, causing it to be a factor in sexual selection. There is very little information to confirm either assumption.

Most women would like to have full-bodied, firm, and well-formed breasts, having learned in childhood that breasts are a sexual attraction so potent that exposing them in public, in print, or on television is virtually forbidden in most circles in Western societies. There are acceptable places for it, such as selected beaches and night clubs. Plastic surgeons came to the rescue of women, but they have not completely ignored men. Perceptive physicians know that there is probably more secret anxiety among men about the size of their penises than among women about the size of their breasts. Surgery to improve the male appendage is simple or difficult depending on the objective, and of dubious value. But there are intrepid surgeons ready to offer their services. Plastic surgery can add length, width, or both. Width is added by transplanting fat from another part of the body, usually by injection. The procedure is called circumferential autologous penile engorgement (CAPE). About $\frac{3}{16}$ to $\frac{1}{4}$ inch of fat can be injected under the skin, which will add roughly from $\frac{1}{2}$ inch to a maximum of 1 inch to the circumference, equal to about $\frac{1}{3}$ inch added to the diameter. Unfortunately, the fat may tend to clump or ball up, or it can be reabsorbed into the system, requiring the operation to be repeated every few months. Adding to the length of the penis is more difficult. For this it is necessary to cut the suspensory ligament that attaches the penis to the body, a difficult operation with risk of permanent injury.

Besides a penis that is gargantuan compared to that of the other higher primates, the human male is endowed with a visible pendulous scrotum and prominent pubic hair, all presumably designed to draw attention to the genitalia instead of concealing them as in the great apes. This has led to the belief that the evolution of the human male genitalia evolved under the influence of sexual selection. But the evolutionary significance of the size of the human penis remains unanswered. It is conceivable that there would have been an evolutionary advantage to a long penis after ancestral humans started rolling under fur-lined skins, or in the grass, and found it more convenient or more stimulating to either the male or female, or both, to use the awkward missionary position instead of the easy-access thrust from the rear used preferentially by the other primates.

Sexologists claim that the ventral–ventral position evolved as the human norm because the thrusting action of the penis against the clitoris in this position is more stimulating to the female. Zoologist Desmond Morris, in *The Naked Ape*, popularized the idea that humans are the most highly sexed primate, and that this quality is related to the frontal signals and sensuality of breasts, male genitals, and pubic hair, and the fact that they have moved forward concurrently with adoption of the preferred human frontal position.

The penis was an innovation that greatly improved the chance of a terrestrial male donor depositing his sperm where it would be most efficiently utilized. It has been retained as a utilitarian appendage throughout evolutionary history since its invention in modern form by the ancestors of present-day reptiles. Some terrestrial animals have succeeded remarkably well without it—for instance, the birds—but they are the exception. In mammals, the penis and associated genital characteristics have taken on an added function—a prominent role in sexual selection—resulting in modifications that make the genitals attractive to the opposite sex and causing, through natural selection, changes to take place in their size, shape, color, and location. Nowhere is evolutionary influence on development of the genitals more clearly demonstrated than in the human species, the "sexiest" of all the primates.

Chapter 8

Big Connections

Surrounded as we are by stories of extinct dinosaurs, it is not surprising that most people seem unaware of the fact that we live even today in an age of giants. Elephants are the largest animals that ever lived on land except for an extinct rhinoceros and the largest dinosaurs, most of which, in fact, may have been aquatic or swamp dwellers. Present-day whales are the largest animals that ever lived on land or in the sea.

To the casual observer, elephants and whales may seem to have little in common except their size. Even their evolutionary histories converge at a time so remote that a common ancestor is hard to define, other than to say that it was a terrestrial mammal, for whalelike creatures took to the sea long after their ancestors had already roamed over dry land in search of a future. Even though elephants chose the land and whales the sea, it is evident that they have much in common. Both are air-breathing mammals that suckle and care for their young, and both are social animals that display cooperation in times of emergency. To be sure, they diverged greatly in feeding habits; elephants are herbivores, whereas whales are carnivores. As luck would have it, both possess attributes that are valuable in commerce, putting them at risk of extinction at the hands of the most vicious and unrelenting predator that ever roamed the earth, *Homo sapiens*. In the end, their salvation may be the fact that evolution has treated them kindly, if one can say that about "red in tooth and claw" survival of the fittest, by rewarding them with uncommonly high intelligence, sensitivity, admiration from a large segment of humanity, and above all, as a heritage of the sex imperative, surprising reproductive capability and adaptability for such large and highly specialized animals.

There is something appealing about big animals that borders on mystique, and this is enhanced by the fact that observation of their private lives is not easy, so reliable reports are hard to come by. Elephants and whales are surviving descendents of Pleistocene giants in a world that is increasingly hostile to their welfare, yet they stubbornly hang in there despite almost unbearable human encroachment. After centuries of observation and exploitation of elephants and whales, much about the intimate lives of these intelligent inhabitants of the earth and the sea remains a mystery.

Pachyderm Passion

The elephants are descendents of a pig-size animal, *Palaeomastodon*, that lived in the swamps of subtropical Africa 40 million years ago. Descendents of this first known proboscidean (long-nosed animal) spread to Eurasia, and from there over land bridges to every continent except Australia and Antarctica. By the Pleistocene epoch, which ended 10,000 years ago, they had evolved into the Ice Age mammoth, the mastodon, the stegodon, and the immediate ancestors of today's Asian and African elephants (*Elephas maximus*, and *Loxodonta africana*).

The huge pachyderms look clumsy, but appearances are deceptive. In addition to great strength, they are gifted with agility, physical coordination, and intellectual perception. They are called pachyderms from the Greek *pachys* and *derma*, meaning literally *thick-skinned*. But this is not to say they are insensitive. On the contrary, the behavior of elephants is one of remarkable sensitivity. Richard Carrington in his book *Elephants* (1959), told how in India he saw elephants show a degree of solicitude for their mates that could serve as an example to humans. And Howard Williams described in *Elephant Bill* (1950) how two elephants became attracted to each other, courted, and finally mated with a fondness for each other that gave no hint of the raw animal lust supposed to motivate the reproductive process in the wild.

The structure of the huge animals shows unique evolutionary anatomical accommodations needed for successful mating. In the female, the opening of the vagina is not near the anus as in other large mammals, but is the termination of a long extension within the body called the vestibule that leads downward to the lower abdomen and opens where the penis and testicles would be if it were a male. The mammary glands are not in the region of the hindquarters as in bovine, equine, and other large mammals, but are forward just behind the forelegs.

The male's testicles do not hang in a bag or scrotum outside the body, but lie within the abdomen and are not visible from the outside. The penis is in a sheath in the normal position along the abdomen. The elephant penis is the largest in the animal kingdom, except for that of the largest whales, and of necessity is very long because the female's vulva is located so far forward. Penetration is accomplished by means of a unique adaptation. When the penis is erect, special muscles that flex independent of the pelvis hold it in the shape of an S, which enables the bull to hook it into the female's vulva.

Ancient naturalists, apparently influenced by the unusual anatomical features of the females, believed that elephants mated face to face as in the conventional human position. Probably few, if any, people had actually seen the elephants mating, and it is not surprising that knowledge of the location of the female's vulva and mammary glands would give rise to the myth ascribing human qualities to the animals, because this was held as evidence of the elephant's intelligence and wisdom. It remained until recent times something of a puzzle as to how the huge creatures managed to copulate.

Like almost all other mammals, the female comes in heat, or estrus as biologists like to say, seasonally. In Asiatic elephants this normally happens between December and February. She has several peak episodes lasting 3 or 4 days, interspersed with short, relatively inactive periods. During peak periods, she is more excitable and shows what one authority described as an "enthusiastic readiness" for the male. She may make characteristic sounds that, among other signs, are apparently recognized by the male.

Elephants are spectacular animals, and no doubt much that has been said and written about them is laced with imaginative and romantic embellishments. But the main features of elephant mating have been corroborated by detailed accounts of reliable observers. Some of the most appealing studies were made of the Asiatic elephant, about which Richard Carrington gave a credible description that probably conforms closely to the behavior of the African species as well.

Before serious courtship actually gets under way, there develops a friendship between the male and female in a sort of "boy meets girl" experience. They begin to keep company in a way that suggests they are "going steady." They are still just good friends, but even though the female has not yet come in heat, there seems to be a bond that goes beyond a purely platonic relationship. They will engage in frivolous contact by playfully pushing one another. They may pinch each other with their mouths or fondle one another with their trunks. Periodically, they break off from these playful episodes of amorous teasing and calmly resume their regular elephant business of eating and looking for food as though nothing special had happened. This on-and-off play may go on for some time. Then, with dramatic suddenness, they mate.

The onset of estrus in the female is accompanied by physiological changes that heighten the sexual drive in both the female and the male. He responds to the changes in her metabolism and behavior with excitement and anticipation. Body odors may play a role. Whatever the subtle means of communication, fooling around turns into serious courtship. The female, unlike the coy girl elephant of her former self, becomes aggressively sexual. She makes gestures and body movements seemingly designed to arouse the male's desire. She may make alternative advances and retreats, provocative movements of her body, and erotic fondling with her trunk. Her teasing is sophisticated and direct, justifying Richard Carrington's description of her as the "Cleopatra of the animal world."

Contrary to the myth held by ancient naturalists, the male mounts the female from behind. Instead of clasping her with his forelegs as some other quadrupeds do, he extends them along the top of her back as far as her shoulders. During this part of the action, there is more or less squeaking, grunting, and trumpeting. When the male reaches the point of making penetration, he slides back nearly to a squatting position, then slowly raises himself to nearly an upright position with his forelegs resting on the female's hindquarters. Intromission is over quickly, sometimes within 20 seconds, and generally lasting no longer than a half minute.

Copulation may be repeated two or three times within a few hours. It is not known for certain how long the relationship lasts. One opinion is that the couple may continue to keep company for several months, during which they will take occasions to fade into the forest at dusk for a tryst. But sooner or later there comes a time when the female has had enough, and she realizes that she must turn her attention to motherhood. Without ceremony or sentiment, she gives her partner the cold shoulder. He will be of no further use to her. He may be a competent lover, but he is no good with kids and shows no sign of being thrilled by the prospect of a family. He may, in fact, by now be looking for another mate.

But the expectant mother is not without friends. An elephant herd is a well-ordered society, and the pregnant female will have the protection and care of the social organization. She seeks out the companionship of another female who will serve as a special helper or "auntie" that will stay with her constantly throughout her pregnancy for help and sympathy and later to help care for the calf. The two will protect the newborn calf by keeping it between them at all times. Without this protection, the calf would be vulnerable to attack by predators. The loss of newborn calves to tigers in India is said to be about 25%. In Africa, the loss to lions is probably less because lions ordinarily feed on other game.

The gestation period for elephants—the time from conception to birth—is 20–22 months. Therefore, a female theoretically could give birth to a calf every 2 years, but normally she will breed only once every 3 years (in general, large animals have longer life spans, longer gestation periods, and longer periods of infancy, than small animals). A female elephant may first come in heat when about 10 years old, but the average age of maturity is about 14 years. Because pregnancy lasts 20–22 months, the average age for a mother to have her first calf is about 16 years. Elephants may live for 60 years or more, but they begin to show old age at about 40; thus, the reproductive potential of elephants is not great. A healthy female elephant could have a dozen or more calves during her reproductive lifetime. Some authorities say as many as 20, but that number would certainly be unusual.

An emotional condition called musth (pronounced "must"), believed to be related to sexual behavior, is a frequent occurrence in bull Asian elephants and in some females. Between the elephant's eye and ear there is a gland, called the temporal gland, lying just beneath the skin and opening to the surface through a slitlike orifice. After the male reaches maturity, the temporal glands periodically become inflamed. The temples puff up, and the glands exude a dark, oily, strong-smelling fluid that runs down and stains much of the face. The bulls usually come on musth once a year, sometimes more often.

Elephants on musth are highly unpredictable in their behavior. They become sullen and moody or suffer from a kind of madness in which they become excitable and violent, sometimes extremely dangerous. Asian elephants on musth have been known to kill a mahout (driver or keeper) who had cared for them for years. A confusing feature of elephant behavior is that an elephant can have an active temporal gland without being on musth, but all elephants on musth show the characteristic temporal gland activity. A bull elephants does not achieve his full stature until about 25 years of age, but he will come on musth for the first time at about the age of 15, although he is sexually capable several years earlier. Like humans, the male elephant is able to mate and reproduce long before he is full grown.

Although musth in domesticated elephants has been studied extensively, there is little known about the behavior of elephants in the wild. The temporal glands of African elephants secrete copiously regardless of age, and they do not seem to show the behavioral aspects of musth so characteristic of their Asian relatives. Adult male elephants in Sri Lanka come on musth regularly once a year, but it can be brought on more often by repeatedly exposing the bull to estrus females. There is a large increase in blood testosterone—the principal male hormone—and this may be associated with the notoriously aggressive behavior of a bull elephant in musth. Urine dribbles almost continuously from the penis when the elephant is in full musth.

When a bull on musth is placed with a female, the outcome is unpredictable. If the female is in heat, he will almost certainly copulate with her, in which case the stress from whatever source seems to be relieved. He grows calmer, and his madness subsides. If the female is not in heat, he may either totally ignore her or violently attack her without provocation. Most authorities believe that musth has sexual significance, but just how it relates to the female's estrus cycle is not clear. Observers report that copulation is more frequent when a male is not on musth than when he is. The exact relationship between the temporal gland and the elephant's sexual activity and behavior remains a mystery.

Behavior of the African elephant differs in important ways from descriptions of the Asiatic or so-called Indian elephant. Iain Douglas-Hamilton described the growth and development from birth of a young male African elephant he named N'Dume, the word for *male* in Swahili. When N'Dume's mother gave birth to him, she already had three calves following her. At first, the baby elephant was nourished only by his mother's milk, but by the end of the first month he was experimenting with grown-up food by pulling out clumps of grass with his mouth. He did not yet know how to use his trunk. From the start he was playful, aggressively attacking and ramming his older brothers or sisters who tolerated him good-naturedly and allowed him to do whatever he pleased without disciplining him. By the age of 5 months, he had become more assertive, and apparently somewhat more obnoxious, climbing over and on top of his older siblings when they lay down to take a nap in the middle of the day. At 6 months he had lost his baby looks, and took on more typical elephant proportions, minus tusks. His gender was not evident at first glance because loose skin surrounded his penis, but his male character developed rapidly, and before he was 1 year old, he was trying to mount other calves, the excitement causing his penis to become evident. This is the first real difference between the sexes to appear according to Douglas-Hamilton. He had never seen a female calf mount a male in such fashion. The young bull's infantile sex play appeared some 13 years before there could be any true sexual relationship. Douglas-Hamilton noticed traces of the musth secretion beginning to appear in another young African elephant when the calf was 1 year old. The excretion appeared only intermittently, but he noticed that the occurrence increased in frequency as the calf grew older.

At about 11 to 13 years, as adolescence approaches, further differences in behavior between males and females begin to surface. The young female becomes bored with rough and tumble fighting and tends to show a maternal interest in the younger calves, apparently in preparation for becoming a sort of teenage baby sitter or "nanny." The bull calf remains vigorously active, engaging in bouts of playful fighting, alternating with episodes of mounting other male and female calves. Eventually he will decide to seek his independence by walking into the bush, leaving the family group behind him forever as a mere episode in his youthful past. A young bull may be forced to leave the family unwillingly. He may become so aggressively bumptious and sexually obsessed that the older elephants cannot stand him any longer and by a series of attacks convince him that he is no longer wanted. But he does not sever his ties completely, for he remains sociable and friendly even though he shows no permanent attachment to the family unit. He will remain in the neighborhood of other elephants, usually not more than a mile or so from another bull or family group.

When a bull encounters a family group, he will sniff each cow to determine whether she is receptive. Douglas-Hamilton never saw any appreciable courtship among the wild elephants he studied in the Manyara of East Africa. He saw a female in estrus being followed by 10 males, but he never saw any serious fights between fully adult bulls. Although the cow's estrus period lasts 2 or 3 weeks, she is actually receptive for only about 3 days, which is a short time in the life of an elephant. The bulls cannot afford to "horse around," or perpetuation of the species might be in jeopardy. If the cow becomes pregnant, she will not be ready for sex again for 3 or 4 years. In the presence of an estrus cow, the elephants become excited and noisy. The bulls exude a secretion from the penis that may continue for several days. On one occasion, Douglas-Hamilton saw a cow emerging from the bush being chased by a large bull with his penis extended and dragging on the ground.

An example of the affection and solicitude that members of a family of elephants can have for each other was described by Douglas-Hamilton, who was told of the incident by an eyewitness. An old cow became ill and died. When she fell, the other elephants in her family group gathered around and tried to raise her to her feet. This went on for several hours without any response from the fallen cow. Then a bull, who happened to be with the group at the time, drove the others off and tried to help the downed cow by himself, without success. Observers have noted many instances of behavior among elephants that express feelings comparable to what we call grief when a member dies. They have been seen to bury their dead under branches, and if a body is decomposed or the flesh eaten when they find it, they may pick up the bones and tusks from the remains and carry them away. Charles Darwin, one of the keenest observers of wildlife behavior, wrote extensively on how animals express their emotions. Unfortunately, we do not know how to make scientific measurements of such emotions in animals except in a very limited way.

Another age of extinction for giant animals of the earth may be underway. The fiercest predator the world has ever known, *Homo sapiens*, is decimating the elephants for their ivory, while raiding their habitat for timber and land, making it difficult for them to hold their own and forcing the elephants, themselves, to ravage their ever-shrinking environment. Despite the elephant's success in perpetuating its species, reproduction may not prove to be enough. "The story of the African elephant could have a happy ending," wrote Oria Douglas-Hamilton, "but the ultimate choice is ours."

A Whale of a Life

The blue whale, *Balaenoptera musculus*, the biggest of all whales, is the largest animal that ever lived. Even the bulkiest dinosaur, *Brachiosaurus*, which may have weighed 50 tons, was a runt by comparison. The longest dinosaur, *Diplodocus*, which reached nearly 100 feet, is exceeded in both length and bulk by the largest whales. The blue whale grows to about 100 feet and may weigh more than 135 tons. An 89-foot monster on record weighed in at 136.4 metric tons (more than 300,000 lb), equal to the weight of four *Brontosauri*, 30 elephants, 200 cows, or 1600 humans.

All whales are air-breathing, warm-blooded mammals that suckle their young. The ancestors of present-day whales branched off from other placental animals very early. The earliest placentals are believed by some authorities to have been arboreal, and if that is true the whales have come a long way—from the trees to the ground to the sea. During that evolutionary epic, they went through great changes to a fishlike shape and other adaptations for a carnivorous life in the sea.

The life and behavior of the gray whale, *Eschrichtius robustus*, is the best known of the great whales. Every year, they migrate from their feeding waters in the Arctic to their trysting place in the warm waters of Baja California, where they mate and give birth in the privacy of isolated lagoons. The lagoons are so well hidden that whalers sailed past them for years without the faintest idea of their existence until 1851, when Charles Melville Scammon, captain of the brig *May Helen*, saw numerous spouts from behind the desert sand dunes that line the coast. He decided to take a look and discovered a narrow channel into what is now known as Scammon's Lagoon, or Laguna Ojo de Liebre (eye of the rabbit), where there were hundreds of spouting whales. Scammon came on it during the peak of the breeding and calving season, which usually extends from January to March in the protected bays and lagoons along Mexico's Pacific shores. He worked the lagoon for several years and, as the most successful whaler of his era, became wealthy from his find. He tried to keep the location a secret but, eventually, the secret got out. Other whalers detected the stench of his whaling station when it was carried to sea by an offshore breeze. Dozens of whalers converged on the narrow channel to intercept the whales. One report said the slaughter was so great that the entire surface of the water was covered with blood.

Whalers arrived every year during the breeding season. At first, they tried the traditional method of harpooning the whales from open whale boats, but they soon found that it was too dangerous. The enraged mothers rammed the boats head-on, and slammed them with their flukes, but the mother whales were handicapped because they had to swim slowly to avoid outdistancing their calves. Before long, the whalers learned the trick of harpooning the calf and towing it to shore, knowing that the mother's maternal instinct would cause it to follow. The calves were left to die.

For the gray whale, it was no contest. By the end of the century, they had disappeared. Twice they were thought to be extinct. Eventually, they reappeared, but they were few. In 1936, when there were probably no more than 100 gray whales, the U.S. Congress passed a law that prohibited killing them. The whales then came back slowly from a precarious brush with extinction. By 1947, there were about 250 whales in the annual migration compared to 25,000 to 50,000 when the slaughter began. The western Pacific race is extinct, but the northeastern Pacific race, sometimes called the California gray whale, slowly recovered by 1975 to an estimated 7500 to 13,000.

The inlets to most of the lagoons used by gray whales as nurseries are so narrow, shallow, and tortuous that it would seem to be a mystery how the whales could find the same lagoon year after year. But many of the whales are half a century old, and they remember the way to their secret trysting places. When the mothers with calves, together with their mates, return north to the Bering Sea, they leave the company of those without calves and take routes that lead through warmer waters. They migrate about 5000 miles each way.

Most of the studies of whale anatomy were conducted aboard whaling vessels as the carcasses were being cut up. This had disadvantages because it required some agility to find one's way around the mass of entrails and blood spread over the deck while the pandemonium of butchering by impatient crewmen was underway. A comparatively small organ weighing a fraction of a ton could get lost in the gore, but a great deal of information was obtained this way.

The male whale is streamlined. The testicles cannot be seen from the outside because they are not contained in an external bag or scrotum, as in most mammals, but are inside the body, well protected behind the intestines near the kidneys and take the form of two elongated organs with white, shiny surfaces. In the blue whale, they can be $2\frac{1}{2}$ feet long and weigh as much as 100 pounds. Though the whale testicles are the largest of all animals, the spermatozoa are no larger than those of a man. The whale's penis is sheathed beneath the skin of the abdomen as in bulls, rams, goats, stags, and other ruminants, and in some species resembles a thin, hard rope. Like the bull's penis, called a pizzle, it was used for flogging when harsh instruments of torture were in fashion. In larger whales, the penis can be as long as 10 feet, and in some whales it has been reported to have a diameter of 1 foot. Roger Scoresby, whose father was a whaler, studied biology and became a whaling captain himself, described in 1820 the penis of the bowhead whale, *Balaena mysticetus*:

> The male organ is a large flexible member and is concealed in a longitudinal groove, the external opening of which is 2 or 3 feet in length. This member, in the dead animal, is 8 or 10 feet in length, and about 6 inches in diameter at the root. It tapers to a point, and is perforated throughout its length by the urethra. . . .

Erection is accomplished mainly by the elasticity of the connective tissue as in bovine bulls and other ruminants, instead of by engorgement with blood as in humans, apes, horses, and carnivores that have penises containing a large amount of spongy tissue that fills with blood during erection. The similarity between whales and the even-toed ungulates in the structure and action of the penis led Everhard Slijper, a professor of vertebrate anatomy at the University of Amsterdam, to speculate that there is a similarity in the way they copulate. He pointed out that bulls, rams, and stags copulate with amazing quickness, the entire act lasting not more than a few seconds. In contrast, stallions take minutes, and carnivores take 15 minutes or more. Bears have been seen to copulate for 45 minutes, and martens (small fur-bearing animals related to weasels and minks) for more than an hour. Professor Slijper's hypothesis is not easy to confirm because it is difficult to observe the mating of sea animals. Most of the baleen* whales sought by whalers breed in warm waters away from the main shipping lanes and in isolated locations not easily detected. What actually happens during mating of the briny giants has been a topic of speculation for centuries. Many of the reports on breeding habits of whales are based on observations made by crew members of whaling vessels. Scientific expeditions have obtained information, though it

---

*Baleen whales, also called whalebone whales, belong to the suborder Mystacoceti, which do not have true teeth. Instead, they have long, elastic, horny, bristly slats, about $\frac{1}{2}$ inch apart that hang down from the upper jaw bone. Their baleens vary in length, depending on the species, from 1 to 14 feet, and from 2 to 8 inches in width. The inside edge has a fringe of hairs that strain small sea animals, called zooplankton, when the whale squeezes out a mouthful of sea water with its tongue. Food consists mostly of krill, a small crustacean resembling shrimp. Gray whales, blue whales, humpbacks, right whales and others are baleen whales in contrast to sperm whales, *Physetes catodon*, the largest of the toothed whales. The whale, Moby Dick, of Herman Melville's classic novel, was a sperm whale, a member of the suborder, Odontoceti, which have true teeth. Sperm whales were highly prized as a source of sperm oil, in demand for use as lamp oil and for candles.

is still meager. An accumulation of reports over more than 100 years gives a rough idea of what takes place.

Before mating, a whale couple may engage in prolonged love play. Humpbacks, *Megaptera novaeangliae*, have been seen caressing each other and slapping each other with their flippers. They may touch gently with their bodies and stroke each other with their flippers as they slowly swim past each other. Scammon wrote that humpbacks would give their partners a playful slap with their long fins that could be heard for miles on quiet days. Humpback whales are notorious acrobats. Observers speak of the thrill of seeing these leviathans leap completely out of the water and fall on their backs with a thunderous splash. Humpbacks are thought to mate face to face as they rise to the surface after coming together with great force. Sexual contact lasts only a few seconds. They are said to dive and swim toward each other at great speed, then rear vertically from the water and copulate belly-to-belly. When they fall back into the water, they make a loud, slapping splash that can be heard from a distance. One pair of humpbacks were seen repeating the performance several times within 3 hours. It is not certain that the couple actually copulate when they jump out of the water. The ritual may be part of their love play. During mating of gray whales in the horizontal position, the pectoral fins protrude from the water while the flukes remain submerged. The act of copulation is reported to last about 30 seconds. In sperm whales (named for the waxlike spermaceti obtained from sperm oil) the act is said to be over within 10 seconds.

Biologist Roger Payne and photographers–divers William Curt-singer and Charles Nicklin, Jr., made a study of the "right" whales,* *Eubalaena glacialis*, at a bay on a bleak and sparsely inhabited stretch of the Patagonia coast where the whales come to mate and nurse their calves. The whales visit the area each season during the winter and

---

*Whalers dubbed this species the "right" whale because it inhabited coastal waters, was slow, easy to catch, floated when dead, and yielded large amounts of oil and baleen.

spring. Mothers and calves were seen to congregate in one part of the study area, which seemed to be set aside by the whales as a nursery. There was another area that was used primarily as mating waters, a sort of "singles bar," where numerous males were always jostling each other and competing for the females. Right whales are completely promiscuous, showing no tendency to form pair bonds. During courtship, a female would be stroked and hugged by males competing for her, and when these overtures were imposed on her to the point of becoming oppressive, she would foil the eager swains by rolling over and lying on her back at the surface. But when she would eventually have to right herself to breathe, the males would converge on her again, dive beneath her swimming upside down, all the while pushing and shoving each other for position. During most of this time, they emitted a complex vocabulary of grunts and groans, but what they meant is a mystery. When a male finally succeeded in getting a female friend alone, they would embrace each other belly to belly with their flippers while mating.

Male sperm whales, *Physeter catodon*, are polygynous. As in most animals that collect harems, the males are larger than the females. The males of fur seals and Weddell seals, which also collect harems, are far larger than the females. But blue whales, like some of the other baleen whales, are monogamous. Jacques-Yves Cousteau and his crew explored one of the Mexican lagoons to study the gray whales' habits. One of the party, Ted Walker, described the gray whales' lovemaking technique. The male and female lie side by side on their backs, then turn to face each other and attempt penetration. The penis is about 5 feet long, far back on the male's body, and sharply curved, like a cane handle. The male is apt to make many tries before hitting the mark, but the lovemaking episode can last for a long time. The Cousteau team saw a pair try to couple time and again, and achieve penetration, or nearly so, several times. When penetration was finally achieved, it lasted only a few seconds, not more than a half minute, but it was repeated every few minutes for an hour and a half.

Most matings of the gray whale probably involve a second male.

They may require three to tango because of their bulk and need for bouyancy. One day Walker sighted an episode in which three whales were "all rolling around together and stirring up a sea of foam." The second male is presumed to help the pair when the actual sexual encounter occurs by placing himself crossways to the coupling pair in order to support them and help them keep their position in the water. Meanwhile, the female uses her flippers to embrace the male. Presumably, the kindhearted helper hopes to receive a similar favor. This kind of sexual activity by a trio of gray whales has been reported by Theodore Walker and several other observers, both in Scammon's Lagoon and in open waters near the California coast during the whales' migration.

Cousteau's crew witnessed the mating of right whales. They were rolling around, first one on top, then the other, all the while caressing each other with their bodies. Sperm whales are said to have more violent encounters. They are dedicated to polygyny. A single male may have a harem of 20 to 50 females and their young gathered around him. Most species of whales are monogamous, at least through a mating season, but gray whales appear to practice polyandry, each female accommodating several lovers.

One of Cousteau's companion explorers, Philippe Diolé, told of sighting a young male gray whale with amorous intentions toward a female with a calf. She wanted none of him and kept butting him with her head to push him away. But the young swain was persistent, and advanced on her with such speed that the force of the collision threw the calf, who had inadvertently got in the way, out of the water. Finally, the angered female slapped the young male hard with her tail and swam away with her calf.

The female whale's ovaries are generally smaller than the male's testes, but in the larger whales they can be as large as a foot long and weigh 22 pounds. In an adult female baleen whale, they resemble a large bunch of grapes, and if one of the "grapes" is cut open, a tiny ovum 0.1–0.2 mm in diameter can be seen, the potential start of an enormous behemoth weighing more than 100 tons. The gestation period—the time from pregnancy to birth—is about 11 months in the migratory whales, corresponding to their feeding and calving habits. Blue whales carry their young for 10–11 months; the fetus of the fin whale, *Balaenoptera physalus*, develops in an estimated 11 months; but the sperm whale carries her baby for 16 months. The fin whale at birth will be 20 feet long; whereas the newborn blue whale will span 25 feet and weigh 2 or 3 tons. Eleven months earlier, it was a microscopic ovum. The young blue whale shows the fastest rate of growth of any member of the animal kingdom. A baby blue whale weighing slightly less than 3 tons at birth will reach a length of 50 feet and a weight of 23 tons 7 months later.

The baby whale is born under water, so little is known about what takes place during labor. Observations on dolphins and other cetaceans in captivity indicate that the young are born tail first, and either swim to the surface within a few seconds or are pushed up by their mothers. The young have little blubber and lack bouyancy until they get their first breath of air. The mother sometimes gets help from another female or "aunt," similar to the way females help with a newborn elephant or hippopotamus. The aunt also may keep watch over the calf while the mother goes in search of food.

Once on the surface, the baby whale has no trouble swimming, and soon finds where to go for lunch. The mother's mammary glands do not protrude like a cow's udder, but are flat, elongated organs extending in the large whales 7 feet in length by 2 feet wide, and bulge slightly. The nipple is in a slit and can sometimes be seen from the outside when the mammary gland is extended. When her calf takes the nipple in its mouth, muscles around the gland contract and pump the milk into the calf's mouth. The milk is creamy white, sometimes with a pink tint, and is 3 or 4 times as concentrated as cow's milk. Those who have tried it say it has a slightly fishy smell and tastes like a mixture of fish, liver, milk of magnesia, and oil. A lactating female is estimated to be able to produce 130 gallons of this rich nourishment daily. No wonder the young blue whale adds 200 pounds of weight every day! The mother nurses her calf for about 7 months. A baby blue at birth weighs slightly under 3 tons; 7 months later it will reach a length of more than 50 feet and a weight of 23 tons.

The suckling calf stays close to its mother. When she comes up to blow, the calf's smaller blow can be seen nearby. William Scoresby, Jr., gave a comprehensive description of the bowhead whale in a classic two-volume work in 1820 and told about the remarkable strength of the female's affection for her calf. She has a strong maternal instinct, and defends her calf with ferocity. She is dangerous when pursued or when her calf is attacked or injured, as many whalers have found to their sorrow.

The violent war between men and whales was dramatized by Herman Melville's story of Captain Ahab of the *Pequod* and the enormous white whale, Moby Dick. But old whalers' romantic notions of hunting down the colossal mammals of the deep fade away in the harsh reality of the slaughter, especially after the adoption of sonar, helicopters, motorized launches, gun-fired harpoons, explosive grenades, and factory ships. Blue whales nearly disappeared from the Northern Hemisphere where only a few hundred remained after more than 100 years of whaling, and the bowhead became commercially extinct by 1900. It is recovering slowly in some areas, but hardly at all in others.

The bowhead, also known as the Greenland or great polar whale, was once so abundant in the polar seas that it was called, simply, the whale. It was slaughtered in large numbers for more than 300 years, beginning in 1611, and became so scarce that the hunt was finally abandoned in the western Arctic in 1914. The bowhead is now protected by international treaty except for Eskimo hunting. Eskimos are permitted to take as many whales as they can, but their methods are wasteful, and it is believed that they lose 5 times as many fatally wounded whales as they capture.

Whales are vulnerable for several reasons. They are slow breeding—usually one calf every 2 or 3 years. The long gestation period, estimated to be 11 months in fin whales, for example, is not conducive to rapid reproduction. Most species have a comparatively short life span for such large animals. Their age can be estimated by the number of wax layers in their earplugs, and by other methods, but none of them is reliable. It is believed that whales live for an average of 25–30 years, but with differences in longevity between species. Calves are susceptible to diseases and predators such as killer whales, sharks, and swordfish. When excessive whaling is superimposed on the other environmental hazards, the whales' limited reproductive potential cannot keep pace with the losses. Even when whaling is discontinued or controlled, they may continue to decline or come back slowly at best.

Elephants and whales are living representatives of an evolutionary adventure that culminated in giant organisms at least 100 million times larger than their single-celled ancestors that emerged from the primordial ooze 4 billion years earlier. Throughout that tortuous climb from the watery cradle of life to dry land—and for whales, back to the sea again—their ancestral genes were reshuffled again and again, and the progeny given a new deal at every step of the way by sexual exchanges of genetic material, which, along with mutations, ultimately enabled elephants to be elephants and whales to be whales. Nature's multimillion-year gamble paid off magnanimously. It is partly up to us to see that the winnings are not lost forever.

Chapter 9

Gender Benders

In Greek mythology, Hermes and Aphrodite had a handsome son named Hermaphroditus who, while taking a bath at Salmacis, aroused the amorous impulses of the nymph of the fountain. He rejected her overtures, but she flung herself on him, embracing him, and appealed to the gods that she might be forever united with him. The gods obliged and the result was a person that was half man and half woman.

Double-sex organisms (hermaphrodites) are common in nature. Although rare in higher animals, hermaphroditism is present as part of the normal reproductive pattern in many of the lower animals. It reaches what is perhaps its greatest sophistication in the familiar brown garden snail, *Helix aspersa*, a native of Europe that was imported to the United States by French and Italian immigrants in hope of establishing a local source of the delicacy known to gourmets as escargot. Unfortunately, the snail's main contribution to the New World was to become a garden pest.

The sluggish snails are far from slow sexually and display the only truly romantic sex life among the lowly crawlers. Each snail has a single reproductive opening, called a genital pore, on the right side of the head. The sex apparatus on the inside consists of complete male and female reproductive systems, plus an embellishment called a dart sac that comes into play during courtship. The genital pore is the opening of a common duct that branches into a vagina, a sac containing a penis, and the dart sac. Far back into the body near the tip of the shell, there is a single gonad, the ovitestis, an organ that produces both eggs and sperm.

When two snails are ready to mate, they face up head to head and shoot their darts into each other. Cupid's arrows! The purpose is a mystery. One theory is that shooting darts is a signal that the shooter is ready. Another theory is that being shot stimulates the partner into sexual activity. After the shooting, each snail inserts its penis into the vagina of its partner and deposits a package of sperm. They may remain connected head to head for a long time. Later, each snail lays one or more batches of eggs in a gelatinous mass in a damp place. They hatch as minute, fully formed snails.

Oysters are notorious for being sexually flexible. These bivalves have simple reproductive organs that make it possible for them to change sex easily. The gonads are neither male nor female but contain follicles (small sacs) lined with tissue that can produce either eggs or sperm depending on age, stage of growth, and environmental conditions. Commonly, an oyster that is male when young will change into a female when it grows older. The European flat oyster switches back and forth from one sex to another throughout its life, and oysters are occasionally found that are producing eggs and sperm at the same time. A change from male to female sex is known by the blockbuster name *protandric hermaphroditism*.

In the worm *Ophrytrocha*, belonging to a group called poly-chaetes, from the Greek *polys*, meaning many, and *chaite*, meaning *hair*, hence *bristly*, the juveniles are always males that change later into females. Strangely, if a couple of mature females meet, one of them extrudes its oocytes (immature ova), the physical appearance of its female jaw changes, and within a week it becomes a full-fledged male with a typical male jaw. After the pair spawn, they reverse roles, with appropriate physical changes. They continue to spawn and reverse sex roles back and forth for as long as they are kept together.

Experimentalists have induced double sex in ribbon worms in the laboratory. Ribbon worms have a long, muscular tube, or proboscis, which they thrust out for catching prey. Because they are supposed to be unerring in their aim, zoologists classified them as nemertine worms, from the Greek *nemertes* meaning *unerring one* (phylum Nemertina or Rhynchocoela). An experimental biologist took male and female ribbon worms, split them down the middle, and grafted them together, male half to female half. Thus he produced surgically a type of hermaphrodite consisting of a single worm formed from the two fused parts. This sort of thing is called a sexual chimera, from the Latin *chimaera*, meaning *monster*. The worm is also called an intersex, with both male and female characteristics. In this case the male half dominates, making the worm in effect a male, at least until the chimera is very old.

The open ocean is populated with small, delicate organisms that float near the surface or are able to swim feebly. They make up the myriad species of floating plants and animals called plankton. Many of them are so delicate and transparent that they are invisible even to most of their enemies. One of the plankton organisms is a class of herbivorous zooplankton (animals) called appendicularians, which are members of the phylum Chordata, small primitive relatives of the vertebrates. A typical appendicularian is shaped like a miniature bent tadpole, with a main body shaped like an egg, and a flexible, muscular tail that is thin and flat resembling a knife blade. Except for one species, they are all hermaphroditic, each individual having reproductive organs of both sexes. Some species produce eggs and sperm from the same gonad, but at different times. They release their sperm into the water through small ducts, but the release of eggs is impossible because there is no pore large enough for them to be extruded. When ready to spawn, the little animals must make the supreme sacrifice. The eggs can be released only when the ovary and body split open, and this results in the animal's death. The eggs are fertilized in the open water. Within a day or two, a young appendicularian hatches from its egg and starts to build an abode of transparent mucus resembling a small balloon equipped with filters to trap phytoplankton on which it feeds. The balloon "house" also keeps its owner afloat and gives protection from enemies.

The vast majority of vertebrates—animals with backbones—are either one sex or the other. An exception is found in a group of tropical fishes called hamlets, *Hypoplectrus*, common around the Bahamas. Each fish is indisputably both sex~s in one, called a *simultaneous hermaphrodite*. The fish are normally solitary, found together only at dusk when they pair up in courtship, and can be seen to enter into a series of spawning clasps. When two hamlets meet in courtship, they act as if they cannot decide who wants to do what. The sex roles may shift back and forth, first one acting the part of the male or the female, then the other. No one knows how they end: whether one extrudes only eggs and the other only milt, or whether both partners shed both eggs and milt.

A popular game fish in the near-shore waters of the Pacific Coast is the sheephead, which starts life as one sex and then changes in the midstream of life. The sheephead is a favorite of scuba divers, but it can be aggravating to anglers because it is adept at nibbling, and it has a talent for hanging around fishing lines to steal bait without getting caught. But if an angler can catch one, it is worth the effort. The sheephead is an ugly fish. Instead of the graceful lines of bonita, yellowtail, or salmon, the sheephead is blocky, with a blunt forehead that slopes down to thick, ugly lips and large canine teeth that protrude from a heavy jaw. A gaudy coloration makes its identity

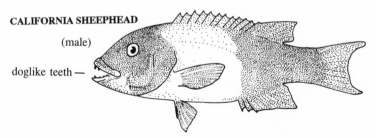

The sheephead, *Pimelometopon pulchrum*, changes sex in the mid-stream of life. All sheepheads are born as females, but most of the old-timers are males. Daniel J. Miller and Robert N. Lea, *Guide to the Coastal Marine Fishes of California*, Fish Bulletin 157, California Department of Fish and Game, 1972.

unmistakable. The male's fins, head, and rear are bluish black, and the middle part of the body varies from red to purple. During the breeding season, a fatty hump develops over the eyes. The female is uniformly colored reddish to rose, sometimes with black blotches.

All sheepheads are born as females, but as they grow older, many of them become males. Most of the old-timers are males, whereas most of the younger mature ones are females. They spawn during the summer at which time the females extrude their eggs, believed to be free floating in the water. It takes 4 or 5 years for sheepheads to reach maturity, and they live for many years. They grow slowly and may reach a length of 3 feet. A sheephead caught in 1956 weighed 36 pounds. Examination of the age rings on the gill covers of a male weighing 29 pounds led to an estimate that it was 53 years old.

A species of fish called *Rivulus marmoratus* that lives in the tidal swamps of Florida was thought at one time to be all females. Biologists were surprised to find that they were actually hermaphrodites with both male and female sex organs in a two-way gonad called an ovitestis. They can produce young by self-fertilization and in this way perpetuate the species even if the population is reduced to the point that the chance is nil that they could find a mate. Self-fertilizing hermaphrodites have been maintained in the laboratory for 27 uniparental generations, covering 14 years. Another peculiarity of this fish is that, under certain conditions, the self-fertilized eggs can develop into true males. Under the strategy of the sex imperative, this serves a biological purpose. Heterosexual reproduction ensures interchange of genes between individuals, with consequent survival value for the species. True females (not hermaphroditic) are either rare or nonexistent, and whether a young fish grows up to be a male depends on the temperature during its development. There are two types of males—a true male with no female ducts, and a hermaphrodite male with the female organs suppressed.

Rivulus is often cannibalistic, eating its own young, eggs, and fellow fish. If only a single fish survives, it can spawn a new population; and even if none remain, the eggs can survive in a damp place for at least a week waiting for an incoming tide. Still another survival strategy of this swamp-inhabiting fish is that it can survive periods of low tide by retreating to damp places under logs or leaves. If disturbed, they squirm and flop their way to a new hiding place and wait for an incoming tide.

Oddball Obsessions

Anglerfish live at great depths in the ocean, but like most deep-sea fish, *Photocorynus spiniceps* has a larval stage that develops in the more food-rich surface waters. When they metamorphose into the adult stage, they descend to the deeper waters where the males and females live different lives. The female develops into a full-size anglerfish, feeding on other organisms on or near the bottom. The male, unlike the female, is very small and incapable of predation. It seeks out one of the females and bites into it, soon becoming permanently attached and parasitic on her. He never grows any larger, but slowly degenerates until he becomes a sperm-filled parasite sharing the female's blood, which circulates through his body as part of her own. This relationship probably has essential survival value for the species. When the female is ready to spawn, there is never the need to search for a male in the dark depths of the ocean, with uncertain results at best. Different species of anglerfish have various ways of angling for prey. In the deep-sea *Ceratias*, there is a luminous tip on a fin, which may serve as an attractant or possibly illuminates the prey.

Ethologists have given a lot of attention to offbeat behavior of domestic and domesticated animals. Konrad Lorenz, in *King Solomon's Ring* (1952), tells about a female barnyard goose that was the only survivor from a brood of six goslings, and consequently grew up with chickens. She fell in love with a Rhode Island Red rooster and became inseparably attached to him. Lorenz, feeling either embarrassed or sorry for his pet, bought the handsomest gander he could find and brought him home as a playmate for the love-smitten goose, but she never gave the gander a tumble. She pestered the rooster constantly with proposals and was so jealous that she did her best to keep him from making love to the hens. Meanwhile, she refused to pay any attention to the gander's ardent advances.

In most mammals, sexual attraction is a response to a long-forgotten ancestral heritage more urgent than things learned from observation or experience. This may be true also for humans, but it does not seem to be true of birds. Lorenz thought that birds reared away from their own kind do not know what species they really belong to. Their sexual desires, as well as other social reactions, are directed toward whoever their companions were during the impressionable period of early life. Lorenz told how he once had a pet jackdaw—a small European crow—named Jock. As the bird came to maturity, "he" turned out to be a female, but nevertheless, fell in love with Lorenz's housemaid. When she married and moved to a village 2 miles away, Jock discovered where she went and followed her, and did not return until the mating season was over. Lorenz raised another jackdaw—this one a male—who showered his sexual attentions on Lorenz himself. A jackdaw pair in nature form a remarkably affectionate relationship. The male expresses his affection by feeding his mate tidbits, bringing chewed up bits of insects and worms and lovingly placing them in her mouth. Lorenz's jackdaw continually wanted to feed him with a choice bird delicacy—chopped worm minced in jackdaw saliva.

The idea of betrothals in the sex lives of animals may seem strange, but that is what actually happens in many species of birds. Small song birds get "married" for only as long as it takes to raise one brood of young. In other species, there is often a long period of courtship before mating actually takes place, and the "marriage" may last a lifetime. Otto and Lilli Koenig followed the lives of young males and females of the bearded tit—a small bird about the size of a sparrow—for several years and found that they became engaged while still juveniles at the age of 2 months, even though they had to wait for another 9 months before they were sexually mature enough to mate for the first time.

Most animals, even if they are not sexually mature, have the courting period late enough to give the male a chance to display his plumage or whatever distinctive features make him attractive to the female. Jackdaws and geese become engaged to a member of the opposite sex in the first spring after they are born, but they cannot mate until they become sexually mature a year later. Marriage is usually for life. Some birds, including jackdaw couples, live together in married bliss for many years because their life span is almost as long as that of humans.

Sex Roles

Females of some species play unique roles. For instance, in some species of ants, females have lost the ability to establish colonies of their own. They must be adopted by established colonies, sometimes of a foreign species, resulting in a mixed colony. In other species there is no working class, so the queen seeks out an alien colony and establishes her own parasitic society, completely dependent on the workers of the alien species for its subsistence.

Some species of ants keep slaves, in which societies the females play a dominant role. The Blood-red Slave-maker, *Formica sanguinea*, an ant common in temperate Europe, periodically raids the nests of other species of ants and carries home larvae and pupae (the stage between larvae and adults). Some of the victims are eaten, but the slaveholders keep the remainder and rear them to adulthood. Because the captives have never known any other home, they remain as cooperative slaves of the ant society. Although the Blood-red Slave-makers can live with or without slaves, an American species, the Shining Amazon, *Polyergus lucidas*, a brilliant red ant, depends completely on slaves for survival. The Amazon workers have lost the ability—or the incentive—to build their own nests, feed themselves, or even take care of their young. But they are vicious fighters—the Amazons of the ant world. The Amazon queen is just as aggressive as her sexless female troops. Moreover, she takes recourse to deceit and violence when she wants to establish a new colony. She hoodwinks a small, weak colony of another species into adopting her, then blandly murders the queen by piercing the enemy's head with her sabrelike jaws.

Lions live in a society that is as well ordered as any in the world of nature. A pride of lions, which is the basic social unit, is a family of usually four or five or more lions. Robert Ardrey described a super pride of 14 lions he saw early one morning in the Congo's Rutshuru valley. Only two were cubs; the remainder were adults or nearly mature juveniles. The members of a pride work together as a hunting unit. Hunting units, if they are to survive, must have strict discipline within the ranks. The leader is the dominant male lion, whose authority is as unquestioned as that of a field marshall, and like a general or an admiral, he seldom makes a kill himself. During the hunt, he moves near the center of the pride, with females quietly spread out on the flanks and in advance. The male makes no pretense of stealth; his terrifying roar sends chills of fear and panic through the prey, and draws their attention away from the lionesses, who do the killing. After the kill, the dominant male is first at the table, not minding that the entire pride waits its turn. Next come those who made the kill— the lionesses. The younger animals get the leavings.

Male mastery reaches its pinnacle in seal societies, which are blatantly polygynous. A dominant male may take on a harem of 10 or more females depending on the species of seal and his success in fighting off ambitious bachelors. A male of the northern fur seal may get and hold a harem of 20–50 females, with an average of about 40, whereas as many as 100 or more females have been counted in harems of the elephant seal.

The master of a seal harem spends all his spare time during the breeding season defending his sovereignty, never taking time to feed during that period but drawing, instead, on his enormous store of blubber for nutrient. The fur seal may go for as long as 2 months without food. The drain on his reserves is severe, for he must not only guard the perimeter of his territory by incessant fighting, but manage during that time to impregnate each female, some as many as four times.

Bloody fights take place, with the winner maintaining the integrity of his domain, and the losers destined to live lonely celibate lives. Protection of the harem needs constant vigilance, especially from forays of ambitious young males. Occasionally an alert male sneaks into the harem while the princely ruler is engaged in combat with another challenger and manages to mate with a female, but if he is caught in time, he will get a thrashing. Severe wounds are inflicted, and if a fight is prolonged, one of the combatants may die from his injuries.

Meanwhile the females recline on the rocks passively unconcerned, resting and sleeping. The female of the northern fur seal is ready for mating within 5 to 7 days after having her pup. She gives it a big meal of rich milk, and then leaves it for a week or so while she goes to sea on a hunting trip that may take her away 100 miles or more. She nurses the pup for 3 months, and then leaves it on its own.

Comparisons between the behavior of birds, carnivores, and other animals with human behavior should be made with reservations. Birds branched off from the reptilian ancestral line at least 150 million years ago, and the reptilianlike lineage that led to humans probably parted much earlier. The first known bird, *Archaeopteryx*, identified by its fossil remains in the shales of Solenofen in Bavaria, Germany, existed during the late Jurassic period, which ended about 135 million years ago. Carnivores and primates both appeared early in the Paleocene epoch, beginning about 65 million years ago, and their ancestors certainly split off much earlier.

Homosexuality

We have seen that there can be considerable flexibility in the choice of sex objects. Homosexuality in various degrees is common among animals, but hardly the same as we know it in humans. Even subhuman primates are not known to have intercourse with partners of the same sex. They may mount or lick each other's genitals, and many animals, juveniles especially, may mount each other indiscriminately under the irresistible compulsion of their newly felt hormones. Usually, as they grow up, their behavior changes under the influence of their maturing physiology, which directs them to pair with appropriate sex partners for reproduction. Only rarely, and probably only under unusual circumstances, do animals in the wild pair to form serious life-long homosexual attachments.

George and Molly Hunt observed lesbianism among pairs of western gulls, *Larus occidentalis*, on islands off the coast of southern California. As many as 8 to 14% of the nesting pairs on Santa Barbara Island are female–female pairs with super clutches of four to six eggs instead of the normal maximum of three. Most of the lesbian seagull's eggs are infertile; the presence of a few fertile eggs indicates that there is some promiscuous mating with males on the island. The homosexual females show some of the same courtship and territorial behavior as heterosexually mated birds, but the female–female pairs do not usually engage in courtship feeding, mounting, or copulating. In heterosexual mating, males normally offer the females large amounts of food during courtship-feeding. This is never done by the females of a lesbian pair. However, a female will occasionally regurgitate food in response to repeated head tossing by her partner in a way that is similar to courtship-feeding by males. One member of a lesbian pair may mount and attempt to copulate with the other, but that is rare.

Homosexual pairing has never been reported in any other group of wild birds, although super clutches have been seen in colonies of the related ring-billed gull, *Larus delawarensis*. Among western gulls, it is apparently a recent development because earlier observations showed that clutches of eggs larger than the normal maximum of three per nest were a rarity. The gulls apparently started the practice around 1968.

Why the seagulls are gay is a mystery. Scientists cannot tell whether it is a behavioral aberration or has an adaptive purpose. It

Brown pelican eggs that never hatched. Interference of the action of the steroid hormones by chlorinated hydrocarbons caused thin eggshells, resulting in crushed eggs. Unusual homosexual behavior of gulls in the same marine environment may be related to similar steroid hormone imbalance due to toxic contaminants.

may be related to the sex ratio of the island population of seagulls. Extra females are allowed to have some promiscuous matings with males so they can pair with other females for nesting and hatch young from a few fertile eggs, whereas without any mate at all, they would not be able to incubate the eggs or raise chicks. This would appear to have the purpose of making it possible for an excess of females to reproduce. Actually, they suffer a reduced nutritional capacity because of the lack of courtship-feeding by the male. Eggs laid by homosexual females are smaller than those laid by heterosexual females, and the chicks that hatch from them are less apt to survive.

The behavior of the gulls may be a result of chemical contamination affecting their endocrine (hormone) systems. Contamination of the food chain by DDT and PCBs (polychlorinated biphenyls) is known to upset the steroid hormone balance in sea birds to the point of causing lethal thinning of eggshells. But there has been little or no study of whether the same or similar hormone changes are related to the apparently new pattern of sexual behavior.

Young adult animals may act more like those of the opposite sex as mating time approaches. A female guinea pig will begin to show sexual arousal at the age of about 45 days when she becomes restless and more active. She will go through the motions of mounting both males and females. This behavior causes the male to become excited, and he in turn makes attempts to mount the receptive female until she assumes a posture that favors copulation. The female remains sexually active for about 8 hours. After that, she kicks at any male who comes near her, and runs away from him.

An explanation of homosexuality is one of the more frustrating tasks of the biological, medical, and sociological sciences. No one can give a reason for it that is satisfying to everyone. An analogy with animals, in which various forms and degrees of homosexuality are commonly seen, explains it no better than animal aggression explains human warfare, or promiscuity among baboons explains human fornication. Let us assume, as the American Psychological Association does, that homosexuality is a normal aspect of sexual behavior. Could it have survival value to the species comparable to the primary sexual function—mixing of genes and perpetuation of the species? If so, it is not apparent, and perhaps this is the source of the dilemma that accompanies a profusion of moral, social, and legal restrictions on sexual behavior. Some types of sexual behavior are clearly outside the range for society's approval, such as rape and torture, or incest and seduction of children. Cash for sexual favors is approved in some cultures and is a crime in others. Homosexual relationships between consenting adults is denounced by some people and defended by others, making homosexuality the most controversial aspect of sexual behavior, and a subject on which there is apparently as much misinformation as reliable data.

The term *homosexuality* refers to participation in sexual behavior with a member of one's own sex. But this simple definition is inadequate because, as a biological phenomenon, there is a wide range of behavior that has homosexual overtones. Among animals, it is common for juveniles to try to mount their brothers and sisters indiscriminately long before they are capable of reproducing. Both their approach and response to sexual advances are experimental and confused. Their glands of internal secretion may not yet be functioning with adult force, nor have they become adults in other respects. Usually a change takes place when the animal achieves adulthood. In adult life, the animal's behavior usually takes on a characteristic pattern that makes its attitude toward sex unmistakable. A male dog, for example, is apt to be aggressively hostile toward a strange male, but congenial and friendly with almost any passing bitch.

Many forms and degrees of same-sex relationships are practiced by people who are predominantly heterosexual, bisexual, or homosexual. Practices differ in different cultures; thus, human sexual behavior is flexible and determined partly by what is fashionable and condoned by society.

Gender is typically distinctive in a way that clearly distinguishes between male and female, but we have seen that in animals gender is not always a simple "him" or "her" designation. Many lower organisms are hermaphroditic. A few are self-fertilizing, although more often than not, elaborate mechanisms prevent it. The apparent value of double sex is that it obviates or minimizes the energy required to find a mate, but it is of dubious value as an evolutionary strategy. Hermaphroditism is rare in vertebrates; fish are the most advanced forms that display it, and it does not occur in mammals except as partial hermaphroditism resulting from natural or experimentally imposed hormone imbalance. The strategy of sex reversal is not always obvious. In the sheephead, a population consisting largely of elderly males would seem to be a poor strategy for survival of the species, but the disadvantage may be more than offset by the predominance of females in the younger crowd.

Chapter 10

The Female World

A Roman legend had it that the city of Rome was founded by twin brothers, Romulus and Remus, who according to the Roman historian Livy, were of virgin birth. Their mother, Silvia, was a Vestal Virgin, and her sons were fathered by Mars, a god of the highest echelon. Even so, Silvia was sentenced to be buried alive for breaking her vows.

The idea of virgin birth is a recurrent theme in the legends, religions, and literature of various cultures. The spread of Christianity and the importance attached to Matthew's statement that the child of Joseph's wife was conceived by the Holy Ghost made the concept of virgin birth both a theological and a biological question that has never been satisfactorily explained to nonbelievers and has never been refuted to the satisfaction of believers. In this mixed mentality of belief and disbelief in what appeared to be a scientific impossibility, a dramatic discovery in 1899–1900 shook the scientific community and astounded the public.

Jacques Loeb, a German-born, French-educated, American physiologist announced while working at the University of Chicago that he had been able to induce eggs of sea urchins to develop and grow without benefit of fertilization by male sperm. Loeb made the embryos develop by placing them briefly in solution of salts that were hypertonic, that is, more concentrated than normal, and then transferring them back to sea water. Scientists were mildly contemptuous, calling his virgin-birth sea urchins "chemical citizens" and "sons of

Madame Sea Urchin." One French periodical, *Annales des Sciences Naturelles*, referred to the virgin-birth sea urchin progeny sarcastically as "Monsieur Chloride of Magnesium." Even Loeb, himself, was at first skeptical of what he called chemical fertilization, and was afraid that he might have been mistaken.

A French biologist, Eugène Bataillon, conducted similar experiments using frog eggs, and was astonished to see fatherless tadpoles squirming about in his culture dishes. He wondered in disbelief if, somehow, frog sperm had been in the tap water of his cultures. Bataillon experimented further, and found that frog eggs could be made to develop into embryos simply be pricking them with a needle. In this way, he stimulated thousands of frog eggs, and many of them developed. He was able to obtain more than 200 tadpoles, and to rear nearly 100 fatherless frogs of both sexes. The chromosome number in the cells of Bataillon's fatherless frogs was diploid, despite the lack of contribution from a male parent. The nuclei of the eggs divided under stimulation and then fused to become diploid, acquiring the same chromosome number as if they had been fertilized. An animal will seldom undergo advanced development unless the diploid number is present.

It is now recognized that in some animals embryogenesis is initiated simply by the mechanical irritation of sperm entry. Eggs of the sea urchin can be entered and activated to start development by spermatozoa from such unrelated organisms as the mussel, *Mytilus*, or the small annelid worm *Adouinia*. Eggs of the toad, *Bufo*, can be activated by spermatozoa from the salamander, *Triton*. In these cases the embryos are haploid; that is, the chromosomes are unpaired. Haploid sea urchins soon degenerate; haploid tadpoles will continue to develop, but with low viability.

It is easy to induce unfertilized frog eggs to start development by pricking them with a fine needle dipped in blood or by exposing them to any of numerous chemical or physical shocks such as a mild electric shock, heat, ultraviolet light, osmotic shock, inorganic salts, organic solvents and acids, agitating them with tiny pieces of glass, or simply shaking them. However, rarely do such fatherless embryos develop into normal tadpoles, and when they survive through metamorphosis, they are usually smaller than normal.

More than two centuries before Loeb and Bataillon made their discoveries, a Dutch amateur microscopist, Anton van Leeuwenhoek, examined all sorts of things through lenses and microscopes that he made himself and was astonished at what he found. One of Leeuwenhoek's observations gave the first description of spermatozoa, and he carefully studied the fertilization of several animals, especially fishes and frogs. Thinking, as Aristotle had believed centuries earlier, that new life comes from the male and that the female contributes merely nourishment and the environment for development and growth, Leeuwenhoek performed various experiments to support his *spermist* convictions. He made a close study of aphids, so-called plant lice, living on black currants. It turned out that aphids were an unlucky choice to support his theory because, to his amazement, he could find no males. He found embryos inside the females in every stage of development, and he watched them give birth to as many as nine young ones in 24 hours. "They bring little aphids into the world without mating with a male," he lamented in 1700. But he tempered his disappointment with the assertion that this sort of thing "is found in no other creature."

A few years later, Charles Bonnet, an amateur biologist, became fascinated with the fatherless aphids. Bonnet, born of a wealthy French family in Switzerland, studied law and received his degree in 1743, but his interest was in natural history. Pursuit of his hobby led to his election to the Royal Society at the age of 23. Looking through his microscope at the transparent bodies of aphids, he could see the young developing inside their mothers. He followed them through their prenatal growth, watched them emerge from their mothers, saw them take nourishment by sucking sap from the tissues of plants they were on, and then in turn, give birth to young themselves without ever having been near males. What Bonnet saw led him to speculate that each female aphid had all future generations enclosed within her body, like an infinite nest of Chinese boxes. The idea was known as *preformation*. Despite his espousal of the bizarre doctrine, Bonnet is credited with confirming the natural occurrence of virgin birth and establishing the fact that production of fatherless offspring is a regular occurrence. A single maiden aphid, if given the proper environment, could through her offspring bring 100 million fatherless little aphids into the world in 1 year.

Birth without benefit of sex in organisms whose equipment is normally designed for sexual reproduction is called parthenogenesis, from the Greek *partheno*, meaning *virgin*, and *genesis*, meaning origin of birth.

The ancient Greeks had insights into the role of eggs and semen that were ahead of their time, but they also had mistaken notions about the details of reproduction. Aristotle observed correctly that drone bees are born without fertilization, but he mistakenly took the queen bee to be the king. Although he noted that the drones leave the hive to soar up into the air, he missed the point of what they were doing up there, calling it a kind of gymnastic exercise. He concluded that fertilization was not necessary in all animals, but that semen played a vital role in animals that copulate. He compared semen to the seeds of plants, and wondered what semen contained that caused it to give rise to new beings when added to the female substance. He called the principle in semen *vital heat* and thought that when this was combined with the secretion of the female, an embryo was formed. Aristotle, along with others of his time, thought that many kinds of animals were impregnated by the wind, and this idea persisted for centuries. There was so much confusion in ancient times about the relationship between sexual intercourse and reproduction that Aristotle tried to convey his theory of sex to people. He discussed unfertilized bird's eggs, which were called hypenemia or zephyria:

> People who say that hypenemia are the remains of former acts of sexual intercourse, are mistaken. Birds such as chickens and geese have often been seen laying such eggs without sexual intercourse. The hypenemia are called by some people zephyria, because it is said that birds receive these winds in the spring.

Domestic fowls were introduced into the Middle East about the time of Nebuchadnezzar (586 BC), and people saw that hens could lay eggs without ever having been with a cock. It seemed to make sense that they had been fertilized by the gentle, warm spring breeze, Zephyrus. Most people did not notice that eggs laid by hens that had not been seduced by a rooster would not hatch but instead would spoil if left to hatch under a sitting hen.

Two opposing theories about reproduction stimulated lively arguments among sixteenth- and seventeenth-century naturalists. A theory favored by William Harvey was that all life originated with an egg. Harvey, an English physician whose fame rests on his discovery of the circulatory system, strangely took some of his views on procreation almost straight out of the book of Aristotle. Born of upperclass parents in 1578, Harvey was studying in Italy at the time Galileo was in his prime. After obtaining his medical degree in 1602 at Padua, he returned to England to set up practice, eventually serving as court physician to both James I and Charles I. Harvey's studies on the evolution of the embryo led him to conclude that the unity of all life could be found in the egg, from which all creatures evolved. Said Harvey, "All animals, even those that produce their young alive, including man himself, are evolved out of the egg." From this came the saying, *omne vivum ex ova* (all life from the egg), an idea called *epigenesis*, that higher animals evolve from the egg by organs being successively formed from the indifferent matter of the egg. The idea was in conflict with the theory of preformation.

The theory of preformation was proposed originally by Jan Swammerdam, a Dutch naturalist whose productive life was greatly shortened by malaria and melancholy during the last half of his life. Although his studies on insects rated him as the father of modern entomology, his greatest fame came from his discovery of blood cells, an accomplishment that was largely ignored for half a century. Swammerdam's studies on embryology led him to believe that all animals evolve in the embryo from previously created parts, a concept directly opposed to the idea of epigenesis. However, the preformation

theory was also contrary to the popular belief in spontaneous generation of lower organisms, a matter of heated controversy until Louis
Pasteur's famous experiments demolished the idea almost 200 years
later.

Some species of animals are exclusively female, males never
having been seen. Is there some evolutionary advantage in this un-
happy state of affairs in which males are not needed and perhaps not
wanted? There is. For one thing, the females can reproduce without
wasting time and energy looking for partners. Besides delays in
finding mates, involving the risk of succumbing to predators and other
hazards, parthenogenesis is the most energy-efficient way of perpetuating the species. If virgin birth becomes an established pattern, it is
possible for it to become the only way, while the male, now useless,
may pass from the scene never to reappear. An example of this state of
affairs is the strawberry root weevil, *Brachyrhinus ovatus*, a species of
female beetle with a short, blunt snout protruding from the front of its
head. No males have ever been found, but year after year, perhaps for
thousands of years, the females lay their eggs among the roots or
crowns of strawberry plants where they hatch into grubs that feed
underground until ready to pupate and emerge as full grown females
with six legs, a snout, and wing covers so tightly sealed that they
cannot fly.

Parthenogenetic insects are apt to be more destructive pests
than those requiring sexual partners because reproduction is more
certain and quicker. Aphids come from fertilized or unfertilized females depending on the season. All summer generations of aphids are
females from unfertilized females. Males finally appear in the fall
when they fertilize only overwintering eggs. However, in some species of aphids, no males have ever been found, in which case generations of females continue to produce young ones indefinitely without
mating.

A remarkable case of sexual and social adaptation is seen in the
corn root aphid, *Anuraphis maidiradicis*, which is assisted in its
attack on corn plants by a totally unrelated species, the cornfield ant,

Lasius alienus. If one examines a field in which these aphids abound, sickly plants will be seen with numerous anthills around them and small brownish ants tunneling among the roots. If the plants are pulled up, there will be found many bluish green aphids about the size of pinheads when full grown, clinging to the roots. The aphids lay their eggs in the fall, then die, leaving only their eggs to spend the winter. The ants collect these eggs and carry them to their own nests where they pile them up but continue to move them from place to place to keep them in an environment having the most favorable temperature and moisture conditions. The aphid eggs begin to hatch early in the spring before there is any corn, but the ants seem to know instinctively that the young aphids must have food, so they carry them tenderly in their jaws to the roots of early plants such as smartweed and grasses that have already sprouted in the field. The young aphids insert their beaks into the sap-laden roots and suck out the juice. They grow quickly. They become full-grown in 2 or 3 weeks—all females—and start giving birth to living young, again all females. Succeeding generations throughout the summer are all females, and like their mothers, producing young by parthenogenesis.

Occasionally winged females will appear and fly to neighboring fields, dispersing the species. But local distribution is done almost entirely with the help of ants. The aphids seem to be helpless without them, for if an aphid is placed on the ground, it acts bewildered, not knowing where to go. But if the aphid is found by an ant, the ant immediately picks it up, carries it underground, and places it on a root. The arrangement is mutually beneficial. The aphids, which have lost the ability to care for themselves, are protected and kept in a plentiful supply of food, while the ants get much of their food from a sweet, sticky, nutritious substance called honeydew that is exuded by the aphids. The ants give the aphids diligent care and seem to know that corn is their favorite food, although the aphids also feed on cotton and grasses. Ants were seen by one observer moving the aphids from a timothy meadow so they could pasture the aphids in a field of corn 156 feet away. The procession included the ants' own young as well as a large number of their livestock. A curious thing happens to the aphids with the approach of winter. Their physiology changes in a way that causes them to give birth to both males and females. The females of the subsequent matings lay dark green, shiny eggs instead of giving birth to living young as their parthenogenetic ancestors did. The eggs are gathered up by the ants and tended in their nests through the winter.

Sometimes virgin birth goes a step further in the direction of efficiency, as with a species of small flies or midges, *Miastor metraloas*, in which even the larvae produce young. Within the ovarioles of the small larvae there develop a number of daughter larvae that break free into the body cavity of the larval parent and devour their mother's tissues. Reproduction by larval or other immature stages is called *paedogenesis*, from the Greek *pais*, meaning *child*. The midge larvae live in the dark depths of holes in rotten tree trunks, and as long as there is plenty of food, high humidity, and constant temperature, the larvae will continue paedogenetic reproduction—larvae giving birth to larvae—indefinitely. But if food becomes scarce, larvae of the next generation become orange, develop structural differences, and become migratory. If they find enough food they become rejuvenated and sexually developed. They will often produce an egg from which a larva hatches that undergoes normal metamorphosis through pupa to adult fly. The same may happen after a sudden change from cold to high temperature. A medium-size larva may produce two to five eggs that develop into adults. Although the paedogenetic larvae are all females, the pupating larvae may be of either sex. During paedogenetic reproduction, there is no reduction division in the nucleus, so the eggs and therefore the paedogenetic larvae are all diploid, having the full complement of chromosome pairs as if the eggs had been fertilized, but with no contribution from a male parent.

A species of female fish that has lost, apparently forever, its male counterpart is the Amazon molly, *Poecilia formosa*. This is a common pet shop fish from Texas and Mexico but buying a pair is impossible because no males have ever been found. The lack of males does not deter the molly from mating. She does it with another species, the sailfin molly, *Poecilia latipinna*, the male of which has no inhibition against taking advantage of the absence of a mate. His spermatozoa penetrate the eggs, but there is no hybridization because the nuclei degenerate and do not make union with that of the egg. However, the eggs become diploid without benefit of a male contribution. Mating, even with another species, is necessary in this case because penetration by spermatozoa is needed to activate the eggs. Carl Hubbs, when at the University of Michigan, found the all-female mollies in northeastern Mexico in the 1930s and called them the Amazon molly after the ladies of Greek legend. The type of reproduction in which sperm penetrate the egg but do not fuse with the nucleus was given the name *gynogenesis*, from the Greek *gyne*, meaning *female*. Gynogenesis is also seen in some nematode worms.

Several species of lizards reproduce without benefit of males. Three species of gecko lizards, which are arboreal, nocturnal insectivorous lizards, *Gehyra variegate*, *Hemidactylus garnotti* and *Lepidodactylus lugubris*, common on islands of the tropical Pacific Ocean, are all females. Whiptail lizards of the genus *Cnemidophorus*, found in arid regions of Arizona and New Mexico, are females that reproduce asexually. According to S. Ohno of the City of Hope National Medical Center, when interspecific hybrids arise, they may escape sterility and establish themselves as all-female species by gynogenesis or true parthenogenesis. Because they are hybrids, they largely bypass the deleterious effects of inbreeding.

Natural parthenogenesis is well known in birds. As many as 80% of unfertilized turkey eggs show signs of embryonic development when incubated, and considerable laboratory success is had with fatherless turkeys, some of which have been raised to adults. All parthenogenetic turkeys have been males. In contrast to mammals, female birds are the heterogametic sex (ZW), whereas the males are ZZ. Parthenogenetic development is initiated by the fusion of two haploid chromosomes, so the sex chromosome complement of the progeny will be either ZZ (normal male) or WW, which is lethal in birds. Thus all turkeys or chickens born without fertilization of the eggs are males. This is unfortunate for poultry farmers, because there would be a great advantage in having an all-female strain of egg-producing chickens.

Some people, among them scientists, wonder whether it is possible to generate offspring without benefit of sex in higher animals, including—perish the thought—humans. Why not? Sexless reproduction in lower organisms is a common occurrence, and experiments with sea urchins and frogs caused the stout walls of male indispensability to tremble and crumble ever so slightly. Still, virgin birth in beetles, bees, aphids, flies, sea urchins, frogs, and even chickens, turkeys and lizards—none of them mammals—seems a far cry from the chance of human females giving birth to truly fatherless children. The possibility seemed comfortably remote when, in 1936, Gregory Pincus managed to stimulate rabbit embryos by subjecting the eggs to chemical or heat treatment. (Pincus gained fame by his studies on reproduction at the Worcester Foundation of Experimental Biology in Shrewsbury, Massachusetts. The studies led to the use of synthetic hormones to prevent pregnancy, a method that came to be known as "the pill.") In 1941, Herbert Shapiro at Hahnemann Medical College caused the egg of a virgin female rabbit to develop by chilling her ovaries with ice packs.

But reports in the press that mammalian embryos could develop to adulthood following artificial stimulation were greatly exaggerated. Ordinarily, an artificially stimulated mammalian embryo will not develop beyond a few cell divisions. In rabbits, it is rare for artificially activated eggs to develop as far as implantation, although activated ova have been reported to develop in mice and rabbits as far as half way through pregnancy. A way of producing offspring that are the exact duplicate of the parent is by *cloning*, a method commonly used in horticulture. The offspring from cloning are called clones, from the Greek *klon*, meaning *twig*. To produce clones in humans or other mammals would require a technique to stimulate parthenogenesis. Human cloning would have advantages over sexual reproduction. A prospective mother in good health and with the capabilities of a genius could have children that would be carbon copies without the risk of introducing undesirable genes from a father. Writer David Rorvik in his controversial book, *In His Image* (1978), described the experience of an anonymous wealthy businessman who arranged to have himself cloned. The procedure was to remove the nucleus from a human egg cell, and replace it with the nucleus taken from a body cell containing a full set of the man's chromosomes. This would be cultured and the resulting embryo implanted in the uterus of a woman. It would probably be necessary to prepare numerous doctored eggs before the procedure could be made to work. Rorvik said it did work, and the publishers of the book said that the author assured them that the account was true. It is safe to say that most scientists were skeptical, and most of them saw it as pure fiction, while acknowledging that such an event might be scientifically possible in the distant future.

Virgin birth in primates is undoubtedly exceedingly rare if it occurs at all. Though theologians may logically differ, there is no *scientific* record of parthenogenesis ever having occurred in primates, nor is there any experimental evidence that it is apt to occur naturally. At Cambridge University, C. R. Austin, an authority on fertilization, wrote in reference to artificial induction of parthenogenesis in mammals, "There are a few reports of the birth of young after experimental induction, but none of these claims stands up to close inspection." Referring to the possibility of virgin birth in humans, Austin opined that there could be a one-in-a-million chance of such an occurrence. He pointed out that the offspring would be female and would resemble its mother closely.

Contrary to a popular cliche, sex is not necessarily here to stay, although it most likely will be a fixture in most species because the survival value is very great, and evolution's pleasure bribe is virtually irresistible. In some animals, actually a minuscule fraction of a percent of higher forms, perpetuation of the species has partially or entirely reverted to reproduction without benefit of sexual fertilization, but species that clone themselves by relying solely on parthenogenesis for reproduction are at risk when subjected to environmental conditions that call for genetic flexibility. Many, probably most if not all of them, will become extinct. The unlikely prospect that the human species may go down the same path to sexual oblivion, whether for evolutionary or pragmatic reasons, is considered by many people a fate too horrible to contemplate.

Chapter 11

Trick or Treat

Many creatures that specialize in night life, or simply live in the dark, can produce their own light. The ability to make light is seen in an array of organisms including microbes, worms, and fish, but for unknown evolutionary reasons, the gift of light making is not shared by the higher animals. One of the things the living flashlights use light for is to signal to prospective sex partners, sometimes so deceitfully that their scheme is a parody, if not a refinement, of what is usually thought of as an exclusively human trait: duplicity and murder for material gain.

Fireflies, also called lightning bugs, actually beetles (Lampyridae), are the world's most successful practitioners of fraud. The females of these spectacular creatures have taillights near the tip of their abdomens that they flash on and off in a coded message to communicate their whereabouts and desires to members of the opposite sex of their species. They also make a career of flirting with males of related species, using the flashing lights to seduce them into a tryst for the sole purpose of eating them. Their light is not the feeble glow one might expect of a small and fragile creature, one-half inch or so in size. It radiates a brilliant luminosity that can be seen at considerable distance. The fact that these little insects can light up at all seems remarkable enough, but even more astonishing is the way in which the firefly flirt uses her flashing "come-on" deceitfully to lure male fireflies of other species to their death.

Fireflies are medium or small insects, active only at night or dusk. Such nocturnal insects are sluggish by day, but at night they put on a spectacular show of flashing lights. Both sexes, as well as the larvae, produce light. In some species the females are wingless, and both they and larvae are called glow-worms. In the Caribbean Islands it is a common practice to enclose live beetles in gauze and wear them as hair ornaments. Fireflies create a spectacular sight when numbers of them congregate around shrubbery or grassy open spaces on warm summer evenings. Their taillights flash a predetermined code giving a message that a firefly of the opposite sex of the same species can recognize. There is no mistaking the message because the male's signal is different from the female's signal, and each is distinctive for its own species of firefly. A male firefly cruising the neighborhood advertises his presence by flashing his code, and if there is a receptive female in the shrubbery she flashes her come-on response. Human romance would no doubt benefit from a communication device as efficient as the firefly's.

Firefly communication serves its purpose well, but there is trouble for cruising males of related species who have no way of knowing about the ability of the "multilingual" females to send signals in codes used by females of their own species. If a male falls for the flimflam he will find himself in an eager embrace, but for a free meal of firefly flesh instead of for sex.

These *femme fatales*, as James Lloyd of the University of Florida in Gainesville called them, have a remarkable repertoire of mating signals. Lloyd, who made a detailed study of the flesh-eating flirts, wrote that he found females of at least 12 species of a single genus (*Photuris*) that mimic the female signal of several species belonging to four different genera. The mating behavior of firefly species involved in this lethal deception is similar. Males fly about emitting a pattern of flashes characteristic of their species. Females, waiting on their perches in the grass or low vegetation, flash their answers in a code characteristic of the females of their own species. The male is recognizable to the female by the duration, frequency, and number of flashes. The female identifies herself by the interval between the male signal and her response as well as by the duration of her flash. The flashes usually last only a fraction of a second, and come at intervals of from a fraction of a second to as long as 1, 2, or 3 seconds. The signal of a common American firefly, *Photinus pyralis*, is typical. The male firefly comes out at dusk and flies around about 2 feet above the ground, giving out a short flash at regular intervals. The female, instead of flying around, prefers to perch on a twig or blade of grass where she waits for the male's signal. When a male flashes within a few feet of her, she does not respond immediately but waits a short time before giving a short flash of recognition. When the male recognizes her signal, he flies in her direction, and they may repeat the dialogue 5 or 10 times until he reaches her side.

The firefly flirt almost always wants to eat after mating and can sometimes hoodwink the first male of another species that she answers. The victim finds out too late that he is going to be a meal instead of a mate, but the intended victims do not always fall for the flimflam. Either her mimicry is not perfect and they detect her "accent" or they get wise to the swindle in some other way before falling into her clutches. Lloyd noted that one female had to answer 12 males of another species before she captured one. Another female answered 20 signaling males without a catch, and then moved to a new perch and had to answer 20 more before she caught one. That kind of wariness is surprising because males of some species of fireflies are so gullible that even if a person flashes an artificial light for about a second, the male will fly toward it. A person with a flashlight can also get a response from a female firefly by mimicking the flashing signal of a male.

What thoughts, if any, go through the head of madame firefly can only be guessed. How does her mate avoid being butchered after mating? Or does he? Females of the genus *Photuris* are carnivorous and cannibalistic. Collectors discover this to their dismay when they try to keep fireflies alive overnight in a jar. Next morning they usually find one female and a few discarded bits and pieces of all the rest. What evolutionary pathway gave a firefly the ability to "speak" several languages? It would be naive to think that the firefly's brain, no bigger than a pinhead, can cunningly plan and carry out a deliberate hoax. Still, as Lloyd states, ". . . the capabilities of the firefly brain are more complex than hitherto suspected."

Males as well as females have a versatile vocabulary. Albert Carlson and Jonathan Copeland, of the State University of New York at Stony Brook, studied the fireflies on Long Island and noted that males switch from their searching code to a courtship code after finding the female. There is a theory, once popular, that the firefly's flash of light serves as protection against enemies, but that is doubtful. One observer, E. N. Harvey, reported that he saw frogs so full of the luminous insects that the light glowed through their bellies. But it is not clear in all cases what purpose the light actually has. The male of the most common Jamaican firefly, *Photinus pallens*, flashes two kinds of signals. When he is stationary he produces a bright flash at one-fourth to one-half second intervals, but when in flight the flashes come only at irregular intervals, and they are so brilliant that they have been likened to a miniature flash bulb, momentarily blinding a person if near one's eyes.

Males of most North American species of fireflies are loners, flying about singly in search of a female. But there are tropical species that gather in trees in great swarms where the males flash on and off in perfect unison, giving a spectacular display of blinking lights. Hugh Smith, an American biologist, told of seeing fireflies in Thailand flashing in exact unison about every two thirds of a second along a tenth of a mile of river front. The most common firefly of Thailand that flashes synchronously, *Pteroptyx malaccae*, flashes about every half second. A New Guinea firefly, *Pteroptyx cribellata*, has a slower rhythm with an interflash period of about a second. Some species have interflash periods of as long as 3 seconds. In some cases, all the fireflies in one tree flash in unison, while in another tree all are flashing in unison but out of step with those in the first tree. Why do they synchronize? No one knows. One theory is that groups of males may be advertising their presence, and that the intensity of flashing in unison is a strategy to attract more females or perhaps females from a greater distance.

The light that fireflies produce is a marvel of chemistry. It is under nervous control, activated by nerve messages from the brain. If the head is removed, flashing stops immediately. Probably hormones are involved, especially the type called neurohormones or neuro-transmitters. It has long been known that if epinephrine (adrenaline) is injected into the insect, it will glow continuously, unable to turn off its light until the effect wears off. The fuel for the light organ is *luciferin*, produced in the insect's body by specialized cells that are abundantly supplied with tracheae, or breathing tubes. The enzyme luciferase, also produced in the insect's body, causes the luciferin to oxidize, giving off light instantly. Oxygen is needed, and this is fed into the system by another chemical substance called adenosine triphosphate (ATP), an energy-releasing compound synthesized and needed by all animals, including humans, for energy-using functions such as mus-cular contraction. Most species have a layer of white reflector material behind the light-producing tissue.

The light produced by firefly luminescence is sometimes called "cold light" because luciferin, unlike wood, oil, or coal, is close to 100% efficient. Practically no heat is given off. The light of American fireflies is typically in the range of maximum visibility (92 to nearly 100%) with practically no heat rays or ultraviolet rays. By contrast, only 2% of the energy in a gas flame is converted to light, the rest given off as heat; only 10% of an electric arc is given off as light, and sunshine is only 35% visible light. Luciferin can be dissolved in water and extracted from the insect's light organ. The light emitted is mostly in the yellow-green part of the spectrum, but there are slight differences in the color of light emitted by various species of fireflies. Dark-active North American fireflies emit green luminescence, and dusk-active species generally emit yellow light. A few organisms produce two colors. One South and Central American beetle larva, *Phrixothrix*, has rows of green lights running along the sides of its body and two bright red lights on its head.

Many kinds of organisms including fungi and bacteria are capable of biological luminescence. The deep-sea angler fish, *Ceratius*, lives at depths of 1500 to 6000 feet where the female preys on other deep-sea organisms. From the upper surface of her body there extends a retractable stalk with a luminous bulb at its tip. The theory is that it serves to attract prey. The angler fish is one of the few deep-sea fish that have functional eyes.

There are many other forms of luminous marine life, and the light given off by some of them is spectacular. As Christopher Columbus approached his first landing in the New World at the island of San Salvador in the Bahamas, he told of seeing lights that resembled moving candles at the surface of the water and promptly claimed to be the first of his ship's company to sight land. Columbus's biographer, Samuel Morrison, thought that the great discoverer was still too far from a landfall to see lights, but that the crew was on edge from seeing other signs of nearby land, and the navigator's imagination played tricks on him. Other people think that Columbus was too good an observer to be easily fooled and that he actually saw lights from swarms of fireworms that he misinterpreted as candle lights.

The ocean-going Bermuda fireworm, *Odontosyllis enopla*, is an annelid worm of the order Polychaeta. Its sex life is synchronized with the phases of the moon. Two or 3 days after a full moon, the females congregate in swarms at the surface of the water, and each worm, glowing with a greenish light, swims around in a tight circle. The display of lights reaches a climax exactly 55 to 56 minutes after sunset. The males, which are only half the size of the females, ordinarily swim well beneath the surface, but now apparently attracted by the circles of light, they swim toward the whirling females while emitting short flashes of light. The entire group now goes into a rotating, glowing circle as the females extrude their eggs and the males discharge semen into the water. The females leave a luminous secretion that has the appearance of a glowing cloud along with the eggs. When the orgy is over, the females continue to glow almost continuously while the males glow with quick intermittent flashes.

There is little doubt that the luminosity of the females is what attracts the males, because if a person directs the beam of a flashlight toward the water during the mating activity, the males will swim toward the beam of light.

A number of species of fish have luminescent capability. So-called flashlight fish of the family Anomalopidae use their ability to create light for multiple purposes: communication, luring prey, avoiding predators, and improving visibility. The four species of fish in the family all have a large organ beneath each eye containing light-emitting bacteria that collectively emit light about equal to the intensity of light from a weak flashlight. The flashlight fish are not often seen because they are small, shy fish that are usually active only on dark nights and in fairly deep water. John McCosker maintained a collection of the fish at the Steinhart Aquarium of the California Academy of Sciences in San Francisco where he studied and observed their behavior experimentally. He told how, after the Arab–Israeli conflict of 1967, midnight patrols along the coast of the Sinai Peninsula saw a faint glowing mass beyond the coral reef. Thinking that they spotted a team of enemy frogmen, the troops set off explosives in the shoals. They succeeded in annihilating the "enemy," resulting in the beach being littered with small, dark fish that continued to glow with pairs of green lamps on their heads.

The light organ of the flashlight fish has a dark backing that protects the fish from being blinded by the light. After surgical removal of a light organ, it will continue to glow for 8 hours. The fish can turn off the light by raising a black curtain over the organ, completely blocking out the light, or it can startle a predator with a sudden flash of light, giving it time to get away. Also, by turning its lights off when it turns, a predator is left confused as to its position. In addition, the flashlight fish has an irregular blinking pattern averaging 2.9 times per minute. Experiments showed that light of the same intensity of that from its light organ was necessary for it to see and catch food. The bacteria that produce the light in the light organs are symbiotic. Attempts to culture or transfer them were unsuccessful, and it remains a mystery how the larval fish acquire them—possibly they are transmitted through the egg.

Fishermen of the Banda Islands of Indonesia learned that the flashlight fish use their lights to attract prey, and that the lights would also attract larger fish, so they adopted the technique of removing light organs and attaching them to their fishing lines about 10 cm above the hook. Other fishermen in Indonesia used live flashlight fish in a perforated tube of bamboo, which gave them reusable lures.

Production of light appears to have been invented independently by several species of organisms, in which it serves different purposes. In fish, light is an aid in attracting, finding, or capturing prey and may serve to confuse predators. In Bermuda fireworms, luminosity sends the message that it is time for mating. In fireflies, light evolved as a means of communication between males and females, which in their nocturnal life, enables them to find each other at mating time. In addition, the firefly's signaling reaches a level of such sophistication that it enables the carnivorous female to use her remarkable linguistic ability for another purpose as well: to lure males of other species to their death, not knowing that she wants a meal instead of a mate. Thus, a strategy apparently designed originally to facilitate the sex imperative, evolved also into a means of providing nourishment for the female and her offspring in the egg stage.

Chapter 12

The Nose Connection

It is common knowledge that animals communicate in secret languages recognized only by themselves. The most widely used language is odor, a chemical language, and in the vocabulary of odors, sexy scents are among the most powerful. Anyone who lives in a country town cannot help being impressed by the extravagant behavior of male dogs when the village bitch, 10 or more blocks away, comes in heat. Fences are jumped, chewed up, or dug under, ropes are shredded, and chains are broken. Love doggedly finds a way.

Naturalists in ancient times seemed to know intuitively that the air could carry sexual signals, although some people got carried away with the idea and thought that animals could be fertilized by the wind. Aristotle wrote of "winds in the spring" that were thought to be capable of impregnating birds. The idea persisted for centuries. The twelfth-century *Beastiary* said about the partridge, "even if a wind blows toward them from the males they become pregnant from the smell."

Body secretions that affect the behavior of other members of its species are called *pheromones*, from the Greek *pherein*, meaning *to carry*, and *hormon*, *to excite*. In nature, scents are valuable, often crucial, means of transmitting messages. We are only vaguely aware of the extent to which silent scents pervade the world of nature where they perform some of the functions that perfumes are supposed to do for humans and much more. Most animals produce secretions that can be smelled or tasted and enable them to communicate as articulately as humans do when using verbal or visual signs of communication. Scents enable plants and animals to communicate chemically, an ability that humans have largely lost or never developed to a high degree of proficiency.

The sense of smell is one of the most delicately tuned sensory systems of the body. It depends on specialized cells, called receptor cells, imbedded in two small patches of membrane high in the nasal cavity. The membrane is called the olfactory epithelium from the Latin *olfacere*, meaning *to smell*. The receptor cells, having hairlike cilia, are elongated inward, forming nerve fibers that lead to nearby extensions of the brain called the *olfactory bulbs* just above the nose, where they connect with nerves leading directly to other parts of the brain, notably the amygdala, a part of the limbic system, the area of the brain most closely related to emotions. The sense of smell detects and identifies chemicals drawn in with the air, or in the case of aquatic animals, with water. To be detected by air-breathing animals, the chemical must be absorbed by the fluid forming the moist surface of the olfactory epithelium.

Most mammals other than primates have a second olfactory organ, called the vomeronasal organ or organ of Jacobson, above the hard palate of the mouth. The organ consists of a pair of blind-ended vessels that open into the nasal cavity or, as in rodents, into the mouth via a pair of fine canals. Nerves carry the stimulus from the receptor cells of the vomeronasal organ to an accessory olfactory bulb lying behind the main olfactory bulb of the brain, and from there to the amygdala. Thus most mammals have an olfactory system separate from the main system. The vomeronasal gland seems to be sensitive to the heavier molecules, sometimes dissolved in liquid as droplets in aerosols. The organ is usually absent in humans, but is present in the embryo, and may be vestigial in some adults.

It is well known that the sense of smell is closely associated with the sense of taste, and that some things that taste good are not detected by the taste buds, but merely smell good. With many animals pheromones are transferred by licking as well as by smelling.

The olfactory receptors in the nose and the olfactory lobe of the brain are well developed in a wide variety of lower and higher vertebrate animals. The membranes in most human noses are blunted and abused compared to the finely tuned olfactory senses of most animals. Still, the human nose at its best can detect at least 10,000 aromas and is sensitive, for example, to as little as 1/25,000,000 milligram of skunk oil (it would take 600 billion times that amount to equal an ounce).

Pheromones regulate courtship, sexual activity, and various social activities in many species of animals. Convention limits the use of the word pheromone to mean chemical exocretions (secreted externally) that affect the behavior of other members of the same species, but a vast array of exocretions communicate information to members of other species. Pheromones in this larger sense include sex attractants, repellents, alarm signals, frightening signals, trail markers, territorial markers, come-hither signals to promote aggregation, growth stimulants, migration inducers, and what appear to be regulators of population density. Nature is full of defensive exocretions of various plants and animals from stinkbugs to skunks. Edward O. Wilson, an entomologist at Harvard University, speculated that there might be civilizations on other worlds that communicate entirely by exchange of chemical substances that are smelled or tasted.

Both sexual behavior and reproductive physiology are dependent on odor cues in many mammals. Rodents are favorites for laboratory experiments on pheromone responses, but field observations on wild animals and experiments on domestic animals confirm the nearly universal dependence on pheromones for normal sexual and reproductive activity. This includes most of the primates.

Pigs, *Sus domestica*, demonstrate the role of pheromones in several reproductive functions. A sow can identify her own offspring shortly after birth by olfactory cues. Puberty in pigs, as in many other mammals, is accelerated by contact with males, and estrous sows will spend more time near boars than will nonestrous sows. If mucus from the vulva of an estrous sow is placed on a dummy, it will sometimes induce a bore to mount it (dummies are often used in collection of semen for artificial insemination). The saliva of a boar contains a substance that excites the sow and promotes normal reproductive development and mating behavior in young sows, called gilts. The substances are probably the steroids 5α-androsterone (a ketone) and the related 3α-androsterol (an alcohol). These are sometimes referred to as 17-ketosteroids. They originate in the testicles and are found in high concentrations in the submaxillary salivary glands of boars, and also in the saliva, fat, and sweat glands. The testicular hormones are responsible for *boar taint*, an unpleasant odor and flavor of pork from uncastrated boars. Either androsterone or androsterol will induce the characteristic mating response of a sow even in the absence of a boar. The active ingredient may be the alcohol steroid because the ketone is mostly metabolized by the sow to the alcohol form.

An aerosol preparation containing 5α-androsterone is marketed as Boar Mate to stimulate mating. It is the only mammalian pheromone synthesized and sold for improving livestock breeding. Androsterone is found in the urine of the human male, although experiments showed that only 50% of the people tested could detect the odor of the chemical.

Scent glands of many mammals are only indirectly related to the action of their sex hormones. Animals as far apart as beavers and bumblebees produce secretions that are used to mark territorial sites. Canids—wolves, dogs, and their relatives—are notoriously active scent markers. In wolves, scent marking is persistently acted out, especially by dominant males, but also by dominant females, and is partly related to the protection of territory.

When female dogs are in estrus, male dogs will spend more time in their presence than when the female is not in heat. Almost every dog owner is aware of this, and it must seem obvious to the casual observer that canid scents convey other information as well. Scats (feces) are powerful sources of pheromone information, some of which comes from the anal sacs that exude secretions on both sides of the anal opening. Vigorous scratching the ground after urinating, seen acted out especially by male dogs and other canids, releases odors from glands in the paws. Unfortunately, we do not know the dog's pheromone vocabulary. Who can fathom what goes on in the brain, especially in the hypothalamus, of man's best friend?

Konrad Lorenz, an Austrian naturalist, related an incident in which his pet bitch recoiled in horror from the smell of a strange dingo pup brought from the zoo. But after the pup stranger spent the night with her own pups in her absence, the puppy scents became so mingled that she accepted the newcomer. A trace of the animosity must have remained, for she later nipped its ear so hard that it never straightened out.

Charles Darwin wrote extensively on sexual selection, especially concerning the efforts of males to excite or allure females. He wrote, "This is probably carried on in some cases by powerful odors emitted by males during the breeding season, the odoriferous glands having been acquired through sexual selection."

Males of most members of the cat family, from tomcats to lions, give off a characteristic odor that is offensive to humans but meaningful to cats. The odorous substance, abundantly evident in the urine that the male sprays around liberally, appears to excite the female. The pheromone in the cat's urine is a fatty substance secreted by special cells in the kidneys under the influence of the male hormone, testosterone. Secretion of the pig's pheromone is believed to be similarly stimulated by the male hormone.

All members of the weasel family (Mustelidae), except sea otters, have anal scent glands: weasels, ferrets, otters, badgers, martins, wolverines, skunks, and minks. The mink, *Mustela vison*, has long been valued because it has one of the most highly prized furs, the popularity of which has declined owing to the sentiment that killing animals for fur is cruel and exploitive. About 75% of mink fur comes from commercial mink ranches. The animal has a well-developed scent gland that produces a strong, acrid, nauseating odor that some people think is worse than the skunk's. But the mink cannot spray its scent as can the skunk, and there is no reason for it to do so. The skunk's scent is for self-defense, but the mink's scent communicates with other minks—a true pheromone.

The chemical identities of several mammalian scents have been thoroughly worked out, but in most cases their complete biological role has not been established. Many of the secretions have an obvious sexual function, but it is also evident that mammals use their odorous secretions for several other purposes as well, such as identification and territorial marking.

The pronghorn, *Antilocapra americana*, commonly called an antelope*, has four kinds of scent glands located in different parts of its body. They serve several social functions. One set of glands in the male just below the auricle of the ear is used for marking, and probably includes a chemical signal for the female during the mating season. The male rubs this gland on vegetation, usually sagebrush, and leaves a mark that is detectable even by the human nose. Other male pronghorns may sniff, lick, thrash the vegetation with their horns then mark the spot themselves. Females react in much the same way. The active ingredient of the pronghorn's subauricular gland is isovaleric acid.

Dramatic effects are seen in mice and some of the related rodents. Chemical signals can dramatically change a rodent's physiology and behavior, as shown by experiments on house mice. W. K. Whitten and co-workers at the Bar Harbor Laboratory in Maine found that male mice exude an estrus-inducing pheromone that is transported in the air. And David Moulton found that because the urine of male mice induces estrus, segregation of the females inhibits it. More surprisingly, the presence of strange mice blocks pregnancy. Volatile external secretions that cause such responses have been called *primer pheromones*, a term used to describe a substance that acts on the neuroendocrine system in a way that produces a slowly developing

*The pronghorn (family Antilocapridae) is only remotely related to African and Asian antelopes, which belong to the family Bovidae.

physiological response. A *signaling pheromone* is one that prompts an immediate behavioral response.

Many species of animals mark their territories with urine, gland secretions, or vaginal smears. This is particularly a practice of rodents such as the Mongolian gerbil, European rabbit, sugar glider, golden hamster, Maxwell's duiker, and the golden marmoset. Leaving a scent serves to define territorial boundaries, identify individuals, families, or populations, and to indicate social status, or even to signal important features of the environment.

The Mongolian gerbil, *Meriones unguiculatus*, is a small, furry, mouselike rodent of northeast China. It moves around by jumping like a little kangaroo. Both male and female gerbils leave an odor in their environment by rubbing an abdominal scent gland on various objects and on each other. Their marking behavior is related to exploration, social dominance, and territoriality in the male. The female's scent gland is smaller than the male's and has to do with exploration and maternal responsibilities. The active ingredient in the pheromone is phenylacetic acid.

Delbert Thiessen, a psychologist at the University of Texas, and his co-workers, made a study of the scent-marking behavior of the Mongolian gerbil. If a male gerbil is castrated, the scent gland rapidly decreases in size, and he reduces his marking activity. Recovery of marking activity is rapid if the male hormone, testosterone, is administered every few days. If he is overdosed with testosterone, he becomes a supermarker, marking about 3 times as often as he would normally. The same effect is obtained if a container of testosterone is implanted in his brain. The castrate will resume marking even though the scent gland may have already withered.

If the ovaries are removed from a female gerbil, scent marking declines. She resumes marking if she is injected with either testosterone or female hormones, but ordinarily it is estrogen and progesterone, the two major female hormones, that control her scent marking. When she is suckling young, the female shows exceedingly high marking activity and aggression; more than normal males. High hormone levels and exaggerated scent marking are linked to defense of the nest and care of her offspring. So scent marking by the gerbil is controlled by the same hormones produced by the gonads that regulate sex and aggression, as it probably is in most of the hundreds of animal species that claim territories with scent markers and defend them by fighting.

Dominant male gerbils are especially active scent markers. In order to convey information to the group by signaling with phero-mone, he must win and control territory. A male that becomes dominant increases his marking, and males that become submissive decrease their marking. In gerbil society, scent marking is a symbol of group and sexual power.

Hamsters are another small animal with a strong link between scent power and sex hormones. The vaginal secretion of the female has a profound effect on the male. He is greatly attracted by the odor, which stimulates sniffing and licking. When the female places her secretion on an inappropriate partner such as a castrated male or an anesthetized male, normal males will attempt to copulate with the male. A male will sniff, lick, and consume the secretion when the female extrudes it during mating.

The sex life of the ubiquitous rat, whose physiology and behavior have many parallels with those of humans, depends on his sense of smell. If a male's olfactory bulbs are surgically removed, his mating behavior deteriorates. Similarly with male mice and male hamsters, if the olfactory bulbs are removed, they will no longer copulate.

The water environment of fish is no barrier to the sense of smell. Odors guide homeward-bound salmon to their ancestral spawning places. Salmon usually return from the open sea to spawn in exactly the same stream in which they hatched. In the belief that young salmon are "imprinted" by some organic substance present in their home streams, experiments were devised in 1965 to check the theory by taking electroencephalographic readings. The olfactory sacs of salmon were infused with water from different sources, and the electrical activity of the brain was measured in each case. When water from nearby sources other than the home pond was used, there was very little change in the electrical activity of the brain, but when water from the salmon's home pond was introduced, there was vigorous response of high amplitude. Practical use is made of imprinting by the Ohio Division of Wildlife whose biologists take coho salmon finger-lings, not native to Lake Erie, and imprint them with the smelly chemical morpholine in hatchery water. When morpholine is dripped into a stream, coho salmon are attracted from far out in Lake Erie to where fishermen wait with baited hooks upstream.

Fish are known to communicate chemically from one fish to another such things as fright, schooling, and recognition of their own brood by parent fish. John Todd and his co-workers found that the yellow bullhead has an acute external taste sense that is useful in searching for food, and it can recognize other individuals of its own species by means of a pheromone. Injury to the skin of some fishes causes the release of a substance that affects the behavior of other members of its species in what is called the "alarm reaction."

Charles Darwin studied a small crustacean, *Tanai*, related to lobsters, crayfish, and crabs. It lives on the bottom of the sea at depths of 12,000 feet. The males are dimorphic, that is, there are two distinct forms. One form has powerful, elongated chelae, or pincers, that he uses to hold on to the female while mating. In the other form, the front antennae have a profusion of peculiar threadlike structures presumed to act as olfactory organs to aid him in locating the females. Darwin called the organs *smelling threads*, structures that are also found in other crustaceans and are always more numerous in the males than in the females.

Among the amphibians, many species of salamanders have a gland in a swelling under the chin, called the hedonic gland, that is believed to produce a secretion that is exciting to the female during courtship. In some species, the hedonic gland, which appears to be a modified mucous gland, is found in other parts of the body such as the tail or groin.

Small Talk

Chemical communication was invented early in the history of life, and is the principal, if not the only, form of communication other than tactile in many species of microorganisms. The presence of chemoreceptors for detecting molecules in the environment and responding by moving toward or away from them, called *chemotaxis*, has long been known in bacteria and other microorganisms. Similar receptors for detection of pheromones are probably as common. *Escherichia coli* has chemoreceptor patches, sometimes localized at the end of the bacterium. The question has been asked, "Does *E. coli* have a nose?" Even sperm may have the equivalent of a sense of smell. British researchers found that the membranes of human sperm contain proteins that are almost identical to those in receptors in the nasal passages of rats. The presence of receptors in sperm suggests that the egg releases an unidentified pheromone, enabling the sperm to find it. The pheromone systems in four species of microorganisms illustrate their importance in the sex lives of microbes.

The common yeast, *Saccharomyces cerevisiae*, widely used in making alcoholic beverages and for baking, consists of a large number of strains useful for various industrial and research purposes. Two mating types produce separate pheromones, and cells of each mating type respond to the pheromone of the opposite mating type. Yeast cells multiply rapidly by asexual budding, but exposure to the pheromone produced by the opposite mating type stops an unbudded cell from dividing and causes it to assume the features of a gamete. There is a visible change in the shape of the cell from its normal spherical contour to an elongated pear-shaped cell, referred to as a "shmoo" shape. The concentration of pheromone determines the amount of shmooing. There then occurs sexual *agglutination*, a term used to describe the behavior of cells of the two mating types when they come together in a mass mating. Despite the orgiastic confusion, the cells manage to pair with their opposite mating types and fuse to form zygotes.

Volvox, you may recall, is a green alga commonly found in temperate fresh-water ponds. There are several species. The typical organism is a transparent spheroidal jelly ball 0.5 to 2 mm in diameter in which there are imbedded an aggregate of 1000 to 10,000 cells, depending on the species. The jelly ball contains both somatic (non-sexual) cells and reproductive cells. Each somatic cell has two flagella that project into the surrounding water and by their whipping motion propel the spheroid in spirited action.

Sexual pheromones have been identified in several species of *Volvox*, and they have a variety of effects. In most cases, the phero-mones are species specific. In *V. rousseletii*, if no pheromone is present, certain of the reproductive cells develop into somatic (non-sexual) cells. In the presence of pheromone, these cells develop into gametes.

Sex pheromones have also been isolated and identified in the ciliated protozoan, *Blepharisma japonicum*. A typical *Blepharisma* cell has one long macronucleus stretching almost the length of the cell and about 15 spherical micronuclei. The organism reproduces both sexually and asexually. In five species of *Blepharisma* studied by Akio Miyake at the University of Munster, Germany, each of the complementary mating types I and II excretes what Miyake calls a *gamone*, a chemical signal that induces conjugation.

Among the more primitive prokaryotic microbes (those without cell nuclei), sex pheromones have been studied in the bacterium *Streptococcus faecalis*. This microbe is a resident of the intestinal tract, and may cause urinary infections. Akinori Suzuki and co-workers in Japan and the United States isolated and identified sev-eral sex pheromones in the bacterium. It has a sex pheromone that promotes the transfer of plasmids from a donor to a recipient cell (plasmids are small bodies within cells that contain inheritable genetic material). Recipient strains excrete several sex pheromones that induce mating by donors that have the appropriate plasmids. Donors having different plasmids respond to different sex phero-mones.

Insect Incense

Even lowly male insects are driven frantic by a female's sex scent in the form of a few molecules carried through the air, sending sensory signals for great distances. Scientists have seized on the male insect's single-minded obsession with the female's sex perfume to lure them into traps where the gullible males are doomed to die a sucker's death.

One molecule of silk moth pheromone is enough to trigger a nerve impulse in one of the receptor cells located in the male moth's antennae. It takes only 30 molecules of the secretion of the female cockroach to drive a male to intense excitement. In one study, a single virgin female of the pine sawfly attracted 11,000 males. A trap baited with 5 micrograms (0.00000018 ounce) of pheromone attracted 400 male Japanese beetles within 2 hours.

Insect incenses can be used to human advantage. One way is to fool the males of injurious species into chasing phantom females, only to find themselves in a death trap. Naturalists have long known that the males of many insects can detect females of their own species at great distances and can locate them even if the females are placed where they are not visible. Collectors found that they could use traps baited with caged females to attract males. Early experiments were conducted by J. H. Fabre (1823–1915), a French entomologist and well-known writer. Fabre had been trying for several years to find the adult moths of a caterpillar that feeds on oak leaves. He finally found a single cocoon sticking to an oak leaf, and put it in a cloth cage near an open window of his house. A female moth emerged from the cocoon, and to Fabre's amazement about 60 males of the elusive moth appeared from their hideaway in the woods and gathered around the cage containing the virgin female.

Fabre did not know what it was about the female that attracted the males, so he set up some experiments to find out. When he put the female in a sealed, transparent container, the males showed no interest in her, even though they could plainly see her. But they were strongly attracted to an open, empty container that the female had previously occupied. The males were attracted even in the presence of strong chemical odors such as naphthalene, hydrogen sulfide, and tobacco smoke. Clearly, some powerful signal undetectable by human senses was calling the males.

At first, Fabre guessed that the female must have been sending out vibrations with her antennae, perhaps supersonic waves that the males received with their own antennae. Sure enough, when he removed the antennae from males, few of them were attracted to the female. It was not until later that Fabre obtained evidence that the virgin female moth exudes a scent, and that it is the sense of smell that guides the male to a female. Thus it was recognized early in the 1890s that male moths were attracted over long distances, sometimes more than a mile, and that a chemical volatilizing from the mature virgin female was responsible for the attraction.

Researchers saw as an early possibility that the females of an insect might be used to lure the males to their death. Indeed, this was tried in 1893 by A. H. Kirkland, an American entomologist, who thought it might be possible to control infestations of the gypsy moth, *Porthetria dispar*, which had arrived in the United States from Europe in 1869 and was causing devastation of trees and shrubs in New England. Kirkland's idea was to use caged females as bait. Marked male gypsy moths can fly as far as $2\frac{1}{3}$ miles, and are drawn to the scent of the female for a distance of nearly one half mile. Kirkland's method showed promise in the early stages of his trials because large numbers of males were enticed into the traps. Unfortunately, the scheme failed because the number of prowling males that remained on the loose— probably chasing uncaged females—was simply too large.

The idea of using female lures to keep pests under control foundered for more than half a century. Scientists, like most people, are not always eager to pursue projects having a history of failure. Only limited use was made of the gypsy moths' sex attractant. A semisynthetic bait was devised by cutting off the abdominal tips of virgin females and extracting the attractant with benzene. The extract was used in traps to locate infestations and to determine the size of the population in an infested area by the numbers of males caught in the traps.

Renewed interest was stimulated by the finding in 1960 that the sex attractant of the silkworm moth, *Bombyx mori*, is a specific chemical compound that could be easily synthesized. This was not an instant discovery. The finding was the culmination of a 30-year search by a team of German scientists at the Max Planck Institute of Biochemistry in Munich, who were able to prepare 12 milligrams of a pure extract from the abdominal tips of 500,000 virgin females (it takes about 28,000 milligrams to equal an ounce). They named the attractant, a long-chain alcohol, Bombykol. It was the first sex attractant to be identified and raised visions of synthesizing similar substances on a large scale to be used for a revolutionary method of pest control. The word pheromone had been proposed by the Germans two years earlier for such substances.

Meanwhile, a team of American scientists was working along similar lines. Martin Jacobson and Morton Beroza, at the U.S. Department of Agriculture laboratory in Beltsville, Maryland, managed to extract 20 milligrams of pure gypsy moth sex attractant from 500,000 virgin female gypsy moths. They identified the compound chemically and succeeded in synthesizing the pheromone. Later, they patented a closely related chemical to which they gave the name Gyplure. One pound of Gyplure is enough to bait 50,000 traps per year for 300 years.

Since then, sex pheromones of a large number of insects have been isolated and identified, and several are used commercially to control or determine the size of insect populations. The sensitivity of insects to sex pheromones is shown by the reaction of cockroaches. In 1976, researchers at the U.S. Department of Agriculture laboratories Gainesville, Florida, obtained a minute amount of sex attractant from female American cockroaches, *Periplaneta americana*, by the laborious process of drawing a stream of air through large milk cans containing thousands of virgin females and collecting the volatile secretion in a trap cooled with dry ice. After nine months of "milking" about 10,000 virgin females, the scientists collected 12 milligrams of pure sex pheromone. The males would display a characteristic response consisting of intense excitement, wing-raising, and attempted copulation when exposed to 30 molecules of pheromone, the amount in 0.00000000000001 microgram, a quantity more conveniently written as 10^{-14} microgram (there are about 28,000,000 micrograms in an ounce).

The role of pheromones in sexual reproduction from bacteria to primates suggests both an early origin of chemical communication and the enduring need for a system of conveying information about suitable mates and sexual receptivity. The chemical vocabulary of different species evolved differently but along parallel lines, ranging from relatively simple organic compounds to complicated steroids. In all, they constitute the most useful system of communication in nature. It serves, among other purposes, to ensure fulfillment of the sex imperative, from single-celled organisms such as bacteria and protozoa to advanced evolutionary forms such as pigs and primates.

Chapter 13

Primate Perfume

Humans have a limited pheromone vocabulary. Compared to more articulate members of the animal community, we are deficient in our ability to communicate eloquently with each other by our body secretions. For example, a dog has 20 times as many nasal receptor cells as a human and much larger olfactory bulbs in the brain.

We take pride in our ability to improve on nature. Millions of buyers and users of perfumes and scented soaps, lotions, creams, powders, and pomades of various kinds are convinced that scents can enhance feminine and masculine appeal. The magnitude of this aspect of human behavior raises multiple questions. Is perfume simply the use of substitute pheromones in the hope of accomplishing what nature failed to provide? Or do natural body odors themselves, through some vestigial instinct, affect sexual behavior of the human primate?

The odor of secretions produced by the female rhesus monkey has a powerful effect on the sexual behavior of the male monkey. This can be demonstrated by inactivating the male's sense of smell with nasal plugs, in which case he will not respond sexually until the nasal plugs are removed. Richard Michael and E. B. Keverne, at the Primate Research Laboratories in Beckenham, England, found odor-producing substances in the vaginal secretion of the females. Six months after removing females' ovaries, when these substances were applied to the female sexual areas, male partners were immediately stimulated to sexual activity. Michael and his co-workers identified the active principle in the female secretion as containing simple organic acids (acetic, propionic, isobutyric, isovaleric, and iso-caproic). They called them collectively *copulins*.

Women produce odorous secretions containing some of the same organic acids found to be olfactory sex stimulants in rhesus monkeys. The output is cyclical and linked directly with the ovulation rhythm. Richard Michael and co-workers working at the Emory University School of Medicine collected vaginal samples on tampons from 50 healthy young women and analyzed them by a method that gave the kinds and quantities of volatile acids present in the secretions. The women produced peak amounts of volatile organic acids during the late follicle stage of the menstrual cycle and declining amounts during the luteal phase (the late follicle stage is just before ovulation, whereas the luteal stage follows ovulation).

The proportions of the different acids secreted by women and female rhesus monkeys were similar except for acetic and propionic acids, which were higher in women. About a third of the women produced only acetic acid. In women taking the pill, the peak did not appear; the secretion of volatile acids remained nearly the same throughout the menstrual cycle. The total amount of volatile acid secretion was also lower in women taking the pill.

It is tempting to think that the volatile secretions of the human female vagina are involved in attraction and sexual stimulation as in higher nonhuman primates. The fact that peak production of vaginal secretions comes during the woman's fertile period supports this view, but there is little evidence that the vaginal secretions are actually attractive to men, although subliminal pheromone stimulation cannot be ruled out. Vaginal organic acids may be vestigial pheromones that once had a dominant role in sexual attraction, as they still do in monkeys, but social conventions of humans may now replace most of the instinctive value of pheromones. If one believes advertisements, genital odors are to be avoided like the plague. Frequent vaginal douching destroys the normal bacterial flora and probably any remnant of presumed ancestral pheromone effect.

The idea that odors play a role in human sexuality is supported by the finding that many of the sex steroid hormones and their breakdown products have detectable odors. More than 30 compounds have been identified in human vaginal secretions. Most of them have detectable odors, but it is not known which of them are responsible for the predominant vaginal odors because they come from a number of vaginal and surrounding sources. Richard Doty and his co-workers at the University of Pennsylvania conducted experiments using human noses—37 men and 41 women—and found that there was a variation in the degree of pleasantness and intensity of the odor from secretions obtained during different phases of the menstrual cycles of women. There were also considerable differences between women, but they found no evidence that the odors were particularly attractive to the men. Whether vaginal odors influence human sexual behavior in some subtle way remains a matter for speculation but of more than passing interest to manufacturers of personal products and perfumes.

The menstrual cycles of women tend to become synchronized when they live together closely. Martha McClintock conducted a study of roommates and close friends at an American women's college. During the academic year, their menstrual cycles progressively converged until they were in many cases occurring at the same time. McClintock thought that pheromones were probably involved. The most likely cues would have been odoriferous compounds in sweat, but vaginal odors are also a possibility. The specific cause of menstrual synchronization remains to be determined.

Odors are an important attribute of taste and flavor of foods and beverages, and a great deal of effort is exerted by specialists to make them more attractive. Nose experts are called on to work such wizardry as making vinyl gloves smell like leather, used cars smell like new ones, artificial flowers smell fragrant, bleu cheese "bleuer," fabrics fresh, strong toiletries unscented, and to put the missing flavor in dietetic foods, to improve the smell and taste of tobacco, and to make pet-repellent odors and pet attractants. But the most intriguing of the scents that appeal to the human nose are the fragrances intended for use on or around the human body.

Advertising blurbs appeal eloquently to unspoken desires: . . .

untamed fragrance . . . mysteriously exciting . . . an irresistible whisper . . . anything can happen when you're wearing . . . long lasting enough to arouse and excite . . . seductive but never obvious . . . to enhance the natural secret feminine scent of your skin . . . nature's fragrance mated to musk . . . the passionate scent . . . put it on your body and it goes to his head . . . so earthy, so sensual, so powerful . . . the fragrance of love and youth . . . the fresh fragrances of all outdoors with the sultry temptation of musk . . . the fragrance of adventure.

Men of course are not left out: . . . expresses a man's individuality . . . the masculine scent . . . the classic tradition of musk . . . the provocative scent that instinctively calms and yet arouses your basic animal desires . . . today's man is more adventurous and expressive than ever before . . . so powerful that it has become the decade's rage.

The National Geographic Society made a survey in 1986 involving over 1.5 million people in several countries. Each person was asked to smell a scratch-and-sniff panel and record their reactions. The survey was devised by scientists at the Monell Chemical Senses Center of Philadelphia, one of the leading institutes of smell and taste. The results showed that women can smell more acutely than men. Surprisingly, pregnant women have a diminished sense of smell, and 1.2% of the participants cannot smell at all, a condition called *anosmia*. There were differences in detecting ability between nationalities, but the reason for the differences is not clear. For example, the percentage of men and women who could not smell the male hormone androsterone was higher in the United States than in other parts of the world. Androsterone is produced by bacteria in human armpits and appears in sweat, especially in men. More than a third (35%) of the U.S. participants failed to detect it. However, it is known that repeated exposure to androsterone improves the ability to detect it. Galaxolide, a synthetic musk, could not be detected by 29% of the participants. Odor blindness for androsterone and galaxolide were related; anosmia for them was paired in a high percent of cases. In the survey, acuteness for odors peaked at about age 20 and declined thereafter, dropping off more sharply after 80, but the drop was less pronounced in women.

The most active scent-producing glands of the human body are small glands in the skin associated with hairs: sebaceous glands and apocrine glands. Humans are nearly hairless compared to anthropoid apes. Desmond Morris called us "the naked ape," but we retain a thick growth of hair in strategic areas, especially the axillae (armpits) and around the genitals. The hair serves as both a sponge and a wick for disseminating odorous substances.

Sebaceous glands produce a thick, oily secretion that gives the hairs and their bases a water-repellant coating. The glands are widespread over the body, with one at the base of almost every follicle except in certain areas such as the upper lip, eyelids, nipples, and surrounding areolae. When the glands occur in those areas, they are called "free" glands. The action of the sebaceous glands is dependent on androgen, so they are relatively inactive until puberty, when they may become engorged with sebum that plugs the pores, causing acne, a condition more frequent in males than in females. The sebaceous gland secretion contains a mixture of fatty substances, one fourth of which is a mixture of a large number of free fatty acids. Most of the fatty acids are odorous or prone to produce odors through oxidation or, probably more important, by bacterial action.

Like sebaceous glands, the apocrine glands are located at the base of hairs, but unlike sebaceous glands (which are widely disseminated over the body), they are restricted in their location. The major sites of the apocrine glands are provided with an abundance of hair, heavily concentrated in the armpits and genital area. Women have 75% more apocrine glands than men, but they may not be as active as those of men. Researchers generally report armpits as having a musklike odor, sometimes reported as resembling the odor of urine, an odor characteristic of some steroids (compounds characteristic of the sex hormones). Eight steroids have been identified in apocrine secretions, one of them the same as a steroid found in urine. More androsterone (a male hormone) was found in the armpits of men than in women. Fresh apocrine excretion has no detectable odor; the odor develops after a few hours, apparently from the action of bacteria. Desmond Morris theorized that the heavy concentration of apocrine glands in the armpits is related to the human adaptation to a frontal approach during sexual contact, which keeps a partner's nose close to the major scent-producing area.

Humans are the smelliest of the higher primates, especially when unwashed. But great effort is exerted, at enormous cost in time and money, to keep ourselves from smelling human. In ancient Egypt, the aristocrats were said to take three baths a day as part of the household routine. Because hair accumulates odors, they did away with it. Women did not let hair grow on any part of their flesh, and men removed the hair from their bodies except for their heads and sometimes chins. Aromatic odors were used to replace the natural body odors that had been washed away in bathing, unlike the practice in later times, as in the unwashed court of Louis XIV, when strong-smelling perfumes were needed to conceal body odors.

The fact is not lost on perfumers that the sense of smell is the most subtle of the five major senses. A provocative scent can evoke a long-forgotten memory or emotional association, much of which may be below the conscious level of the mind. Curiously, some of the more appealing perfumes contain extremely diluted concentrations of odors that by themselves would be disagreeable, even repulsive and disgusting. They are mainly those that are related to biological functions such as body odors of various kinds.

The formulation of ointments, unguents, pomades, oils, and perfumes became a flourishing industry in Roman times. Laws were enacted to distinguish ointments used for health purposes from those intended only for pleasure. Men made liberal use of ointments, applying them to the entire body including the head and beard. It was considered good form to use them before festal dinners, for which purpose, according to Petronius, the host distributed ointments to the guests beforehand.

Ovid, an authority on lovemaking, offered restrained advice to women about odors, "I say don't let your armpits smell . . . but I don't have to tell this to the local women." He warned women against using lanolin, a bland material from sheep's wool that is useful in lotions but noted for becoming rancid and obnoxious. Ovid recommended a variety of fragrant lotions for the complexion, but had only brief advice for men, "Don't let the lord and master of the herd be offensive to the nose."

Important advances in perfumery were made during the Middle Ages. Avicenna (the Latinized name of ibn-Sina), the famous Arabian philosopher–physician of the early eleventh century, was the author of a standard text, *Canon of Medicine*, in use for more than 600 years. During his dabbling with ointments, he discovered the process for distilling volatile oils from flowers. Arabian perfumery became a fine art, and was copied far and wide. The Crusaders were quick to pick it up, and returned to Europe with all the art and skill in Asian perfumery they could acquire.

Because many of the natural materials used in perfumes are rare and expensive, a great deal of effort goes into finding synthetic pheromones that will have the same effects. Chemicals in many of the natural odors have been identified and synthesized, and in other cases new chemical compounds are made to simulate the natural odors. There is hardly any fruit or floral scent that cannot be approximated with synthetic perfumes. The natural scents are almost all complex mixtures of many ingredients, and variation in one of them may completely alter the scent, so many perfumers are content with a mixture of synthetic products that gives practically the same odor as the natural fragrance.

In an ordinary mixture of odorous materials, the more volatile ones will evaporate first, creating a series of different odors over time instead of leaving what may be the desired ensemble of fragrant olfactory impressions. Perfumers retard the evaporation of the more volatile ingredients by adding a material to the mixture that is less volatile than the perfume oils. A material that will do this is called a fixative and is most often selected for its ability to blend with the other odors and complement the main fragrance. Some of the most desirable fixatives are animal secretions that might best be described as basic body odors. The most highly prized ones are presumed to have sexual significance.

Unfortunately—or fortunately, depending on the odor—the olfactory system becomes fatigued quickly. The sensitivity to a strong odor may decline within a few minutes or hours to the point that it is no longer detectable. Thus the sensation of an odor may persist for some time or last only microseconds. Each odor has a threshold known to perfumers. For example, high concentrations of civet give off the intensely disagreeable odor of skatole (skatole is one of the components that gives the disagreeable odor to feces). But when diluted, skatole can be reduced to below its threshold intensity, at which point the awful odor of skatole can no longer be detected, and the warm flowery note of civetone comes through. Whether the skatole continues to be perceived below the conscious level is a provocative question. A look at the sources of some of the animal fixatives may give a clue.

Musk has long been a highly prized constituent of expensive perfumes. The musk deer, *Moschus moschiferus* (family Cervidae), from which musk is obtained is a small, powerfully equipped deer inhabiting forests on the slopes of the Himalayas in Tibet and Manchuria to an altitude of around 8000 feet. It is a heavy-limbed, somewhat awkward-looking animal that stands only 20 inches at the shoulder, about 2 inches higher at the rump, and is almost unique in having no antlers (only one other deer species lacks antlers). The shape and appearance of the musk deer are more suggestive of a gigantic rabbit than a deer.

Musk is produced only by the males and only during the rutting season.* The musk gland is in a pouch in the skin of the belly in the region of the prepuce. The secretion forms in the surrounding tissue,

---

*The word "musk" is derived from the same root as the Persian *mushk* and the Sanskrit *muska*, meaning "testicle."

forming a "musk pod" about the size of a crab apple. The odor is evidently attractive to the female deer. Musk is obtained by cutting out the musk pod and drying the contents. A pod yields about an ounce of musk, which, owing to the increasing rarity of the deer, is worth a lot of money. Heavy hunting reduced the numbers to the point that the musk deer became scarce early in the century.

The odor of musk is disagreeable, but it loses its unpleasant quality after aging in alcohol. The active ingredient, called muskone,* is the most highly prized of the animal fixatives. It will impart body and smoothness to perfumes when diluted so much that the odor of musk is no longer detectable. However, it is also used for its own odor in so-called Oriental perfumes, usually classified as "heavy perfumes."

Civet is another valued perfume ingredient of animal origin. Civet is a soft, fatty secretion from the perineal† glands of the civet cat, *Civettectis civetta*. Collection of the civet cat's secretion probably originated in Ethiopia, although the cat is native to many countries. A grown civet cat stands about 10 inches high and is about 2 to 3 feet long. They are captured and kept in pens constructed so that the hot sun keeps the temperature higher than normal. This combined with constant teasing increases the output of secretion. About every 4 days, the secretion is scooped from the glands with a wooden spoon made for the purpose.

Crude civet, like musk, has a disagreeable odor. One of the ingredients is skatole, but after diluting and aging it, the skatole odor disappears, and the sweet, somewhat floral odor of civetone emerges. Civetone is the main ingredient although a number of other odorous substances are present. Small residual amounts of skatole in the final product are probably helpful because skatole itself is used in perfumery in high dilutions as a fixative, as is a related chemical, indole, also a contributor to the odor of feces. Secretions of several other

---

*A cyclic ketone related chemically to civetone from the civet cat.
†The perineum is the region between the anus and the genitals.

animals have been used extensively for their value in perfumes. The muskrat, *Ondatra zibethica*, a rodent belonging to the same family as rats and mice, gets its name from a pair of scent glands on its lower belly. The odor is attractive to muskrats of the opposite sex during the mating season. To the human nose, the secretion gives off a sweet, overpowering odor that quickly becomes cloying, but in very small amounts, it has a pleasing fragrance. It is said that in former times when toiletries were not easily available, ladies would obtain a delicate perfume by placing a kerchief for a while over the body of an unskinned muskrat. The muskrat secretion, called *musc zibata*, is used to replace or complement the Asiatic musk. It has the unusual property of increasing the characteristic musk odor fiftyfold.

The beaver is almost equal to the capybara of South America as the world's largest rodent. Large specimens will weigh as much as 100 pounds. The beaver was probably more responsible for opening the frontier of North American than all other attractions combined. Beaver hats were the rage, and the profits in beaver pelts were enormous; one trapper was said to have earned $50,000 in a single year. But more profitable than furs, and what in the long run nearly drove the animal to extinction, was the beaver's musk glands. The musk glands, technically the perineal glands, produce a brownish orange exudate called castoreum or castor. In America, most castoreum was obtained from the Canadian beaver, *Castor canadensis*. The secretion became the most widely used animal fixative in perfumery, but the product is no longer available in quantity because the animal was driven to near extinction by trappers and hunters, and it never completely recovered in spite of conservation measures. The European species, *Castor fiber*, was hunted almost to extinction more than 200 years ago.

Ambergris is one of the best known of the perfume fixatives from animal sources, although it is rare and therefore little used. It is a waxy substance obtained by cutting open the stomachs of captured whales, but it is sometimes regurgitated and found floating in the ocean or stranded on the shores of a tropical beach. An alcoholic solution of ambergris has a distinctly musky odor.

Synthetic chemicals have added greatly to the complexity, and perhaps the quality, of perfumes. The question is: do perfumes actually have the seductive qualities their trade names and advertising blurbs imply? Confusion comes from the habit many users have of daubing quantities on the body far above the amounts needed for subliminal effects and splashing on inappropriate odors that may lead to faulty human judgment of the response. In the end, such clumsy efforts might be beneficial anyway. The psychological effect on the user may be more important than any effect on the intended smeller. After all, if people think what they did makes them irresistible, they probably are.

The loss of sensitivity to sex pheromones by humans, compared to the capability of other higher primates, is apparently an evolutionary trend away from the silent communication of chemistry in sexual relations, simultaneously accompanied by a trend in favor of the social conventions of speech, body language, and conduct that left us the ability to fulfill the sex imperative as well as, if not better than, before. Vestiges of the ancestral pheromone attributes may be functioning at the subliminal level, although this has not been proved. Still, people think they can improve on nature by being more attractive than nature allowed. Men and women do not want just to smell *better*, they want to smell *seductive*.

Part III

Sex Organs

Chapter 14

Gonads

About a century ago, people began to think that symptoms of old age in men could be accounted for by a decline in the function of their testicles. A prominent French physician, Brown-Séquard, at the age of 72 thought wistfully of his vanished youth and pondered what might happen if he could rejuvenate his flagging virility. And who would make a better subject for rejuvenation experiments than a 72-year-old man with a basic knowledge of physiology and long experience in observing the effects of medical treatment? *Mais naturellement, le docteur* himself. He determined to do something about it.

Dr. Brown-Séquard made some extracts of guinea pig and dog testicles and injected them into himself. He became convinced that the injections made him gain in vigor. In a report published in 1889, he described in detail the benefits he experienced. The experiment gained wide comment, and there followed half-humorous, half-serious talk of transplanting "monkey glands" in aging men to rejuvenate their declining vigor. There were even reports that such experiments had been performed. If true, they must have been among the first experiments in organ transplants.

Unknown to Dr. Brown-Séquard, his preparations did not contain the male hormones of the testicles. He had used aqueous extracts, which could not have taken up the water-insoluble steroid hormones. His results were an example of the power of suggestion inherent in the well-known placebo effect. It has been known for a long time that a placebo (an inert medication) is sometimes effective in treatments for sexual impotency. Physicians report that injections of saline solution will sometimes get the desired result.

Brown-Séquard's experiments stimulated widespread interest in the magical effects of animal glands. As late as the 1960s, Somerset Maugham was one of many who went to Switzerland for injections of "goat hormones." The novelist managed to live into his 91st year, apparently with no complaints about his virility. In the United States during the 1920s and early 1930s, "Doctor" John R. Brinkley established a thriving practice of supposedly transplanting goat testicles into aging men to rejuvenate their flagging sexual potency. He recruited patients by broadcasting commercials over what was said to be the most powerful radio station in the country. His patients had the privilege of picking their own donor animal from the doctor's herd of frisky goats. Brinkley held an M.D. degree from a diploma mill, an honorary degree from the University of Pavia, Italy, and was licensed to practice medicine in several states.

Brinkley began with a clinic in Milford, Kansas, where he quickly became a millionaire charging a minimum of $750 for a "transplant." The glands of a very young goat would bring a fee of $1,500. The operation took 20 minutes. When Brinkley's radio station license was revoked, he set up a station in Mexico across the Rio Grande from Del Rio. He pretended to perform prostate operations and peddled a concoction containing a blue dye and weak hydrochloric acid. He acquired several cars and yachts and a private airplane that he flew back and forth to Little Rock, Arkansas, where he established a hospital. Alf Landon narrowly defeated Brinkley in one of the three tries he made for governor of Kansas.

Testicles

A massive amount of evidence accumulated over the years shows that there is, indeed, a relationship between the secretions of the testicles* and masculine performance and behavior. The masculinizing effect of the testicles was known thousands of years ago, but the relationship between testicles and the principal male hormones, collectively called androgens, was not known until the middle of the last century when a flowering of experimental biology led to the discovery that the testicles are glands of internal secretion as well as producers of sperm. In 1848, Professor A. A. Berthold, a German zoologist–physician in Göttingen, caponized (castrated) young cocks, and he noted that their combs atrophied. In some of the caponized cockerels he replanted a single testis into the body cavity without any of its original nerve connections. The combs grew again to normal size, and the cockerels exhibited normal sexual behavior. Berthold correctly concluded that the testes release a substance into the blood that causes the male animal to develop its secondary sex characteristics and maintain typical male behavior. We now know this to be true throughout the animal kingdom, and furthermore, that the male hormone has a variety of secondary effects. Even a dog's cocking his leg to urinate is related to the pervasive influence of androgens. Berthold's findings were the first published account of hormone secretions from an endocrine gland. Transplantations of bird testes had been performed as early as 1771 by John Hunter, a British surgeon who also implanted cocks' testes into hens and observed masculinizing effects. He performed various experiments, but apparently it did not occur to him that the effect was from secretions released by the testes.

There is a saying, "a man is only as good as his glands," and insofar as this is true of men, it is also true of women. But it is an expression of partial ignorance because it gives no hint of the extent to which the glands and their juices, the hormones, determine sex and

---

*Testicle is an exact synonym of testis (plural = testes). Many biologist prefer *testis*, probably because it is more appropriate as a generic term for organs in primitive animals that produce sperm, and because *testicle* is derived from the Latin *testiculus*, the diminutive of testis, another meaning of which is "witness."

sexuality, or how the glands interact with the nervous system as well as with each other to form an integrated system of remarkable elegance. Sex glands and hormones are present in primitive organisms, but they have been most thoroughly studied in the higher animals, especially humans, and it is by looking into our own bodies that we can best get a panoramic view and see the powerful influence of the sexual gland–hormone system.

The actions and interactions of the glands and their hormones in controlling growth, sexual development, and behavior are among the most finely tuned life processes to be found in nature. Their powerful action, their timing and purpose, their delicate balance and counterbalance, and their pervasive influence on important tissues and organs of the body give a fascinating glimpse into the mysterious forces that control sexual development and behavior from the instant of conception.

Hormones are substances that control or regulate the functioning of tissues and organs. The name is taken from the Greek *hormaein*, meaning *to arouse or excite*. Most of the hormones are produced in specialized glands of internal secretion, referred to as *endocrine glands*, from the Greek *endon*, meaning *within*. Only minute quantities are needed. The gonads normally produce more hormone than the body can use, so some of the excess is excreted in the urine. It was from urine that the first male hormone was isolated and identified, in 1931, by the German biochemist A. Butenandt, who managed to eke out 15 milligrams (about 0.0005 ounce) of crystalline substance from 15,000 liters (about 4000 gallons) of male urine. The hormone was named *androsterone*. Later, it was found that there was a difference between androsterone and extracts taken from testicles, and that the testicle extracts were several times more powerful than androsterone. The testicular substance was given the name *testosterone*, and this is considered to be the most important male hormone. Testosterone is several times more potent than any of the other androgens except a derivative, dihydrotestosterone, which is the active androgen in at least some target tissues. Many substances having androgenic action have become known; all belong to a class of chemical compounds called *steroids*.

The main action of the male hormone in humans is seen in the changes that transfer a boy into a man. During puberty when these changes are taking place, the young male becomes capable of reproducing, and this capacity is accompanied by modifications in his physique, physiology, and behavior. Before puberty, his relatively small testicles secrete some hormone, apparently enough to suppress by feedback the activity of the pituitary gland in producing gonadotropin, a hormone that stimulates the gonads into action (the pituitary gland and its secretions are discussed later). Then suddenly, the pituitary breaks out of its restraints and starts pouring gonadotropin into the blood stream. The testicles begin to enlarge, the penis and scrotum start to grow, and hair appears around the genitals. At the same time, there is a spurt in growth, and the boy gains height rapidly. The growth-promoting action of androgen causes his skeletal muscles to develop and his bones to thicken, resulting in a pronounced increase in body weight and increases in vigor and physical strength.

His testicles reach adult size before all the other changes of puberty—the so-called secondary sexual characteristics—are completed. His skin becomes thicker, and there is an increase in the sebaceous glands, causing his skin to become oily. The sebaceous glands are apt to become plugged with sebum and infected, causing the embarrassing condition known as acne. Subcutaneous fat tends to disappear, causing the veins under the skin to become more prominent. Hair begins to grow in the armpits and on the trunk and limbs, and eventually forms a pattern typical of the male. He may have a slight development of one or both mammary glands, but almost always this subsides or disappears later. Development of the larynx necessitates a readjustment in the tone of speech, which may cause him difficulty at first but eventually brings about a permanent deepening of the voice.

Beginning early in puberty, the young man starts having frequent erections. His libido may become overpowering, and masturbation may become a regular or irregular pastime. He eventually grows a beard, though much to his anguish, this is delayed longer than the other changes and is the last of the transformations to reach full development. Often at about this time, the young man's spurt in growth ends. Though some growth may continue over the next few years, only slight additional height is possible. Our youth has reached physiological manhood.

On the other hand, if the testicles fail to function, or if they are removed during boyhood, puberty does not take place. A castrated boy will continue to grow, but he will become abnormally tall, with long limbs and exceptionally large hands and feet. In sharp contrast to his stature, he retains the appearance and demeanor of a prepubertal child. Because the larynx does not develop, his voice remains high-pitched, his skin stays soft and thin, it tends to wrinkle around the eyes and mouth, and he may have a yellow pallor due to a combination of deficient melanin, poor circulation, and mild anemia. His skeletal muscles remain underdeveloped, and his lack of musculature is made more evident by a layer of fat that forms beneath the skin. Fat accumulates prominently around the shoulders, breasts, upper thighs, hips, and abdomen, giving an impression of femininity. Facial hair is scarce or absent, there is scant axillary or pubic hair, and the growth of hair over the body is short and fine. His genitals are no further developed than those of a child, and he has no sexual drive. The action of androgen becomes quickly evident when it is given to a boy before puberty or to a eunuchoid man. Effects on the skin can be seen within an hour, erections appear within 1 or 2 days, and there follow effects on development and growth, often continuing in some degree for 2 or 3 years.

Androgens are important in maintaining sexual capability and male behavior. This is demonstrated by the reversal in behavior that takes place when a castrated animal is given male hormone. In this case, an animal previously rendered incapable of mating by castration may again become a capable mating partner—except for the ability to impregnate. Although testosterone is regarded as the principal male hormone, several other androgenic hormones are recognized whose functions are not as well understood. There is evidence that some of these "weak androgens" can be converted to testosterone in the liver and other tissues, and this may be a way of getting high concentrations of testosterone into target tissues. There is some evidence that testosterone itself is not the ultimate male hormone, but is converted to an equally potent substance, dihydrotestosterone, within the cells of the target tissues.

Ovaries

The ovaries—female counterparts of the testicles—secrete estrogens, which produce effects as remarkable as the androgens in males. Estrogens are largely responsible for the changes that take place in girls at puberty, and they have a role in causing the transformations that occur when a girl begins to acquire those intangible qualities called womanhood. Estrogens are the feminizing hormones as androgens are the masculinizing hormones. Estrogens bring about the development of the vagina, uterus, and fallopian tubes, the latter destined to carry the ova to the uterus when they are released by the ovaries. Estrogens cause changes to take place that give the body its characteristic feminine form, including enlargement of the breasts, formation of fatty tissue, shape of the bones, the skeleton, contour of the body, and softening of the skin. Estrogen is responsible for pigmentation of the skin around the nipples and contributes to the growth of hair in the armpits and pubic region (however, the growth of axillary and pubic hair of females is due mainly to androgens produced by the adrenal cortex, another sex gland). On maturity, the configuration of the body becomes characteristically female: narrow shoulders, broad hips, thighs that converge, and female distribution of fat in the breasts and buttocks. In women, the larynx retains the same proportions as before puberty, so the voice does not greatly change but remains higher pitched than men's voices in general. In women there is less body hair than in men, and the pubic hair grows in a flat-topped pattern instead of pointing upward as in men. Finally, the secretion of estrogens is strongly rhythmic, causing a cyclic intensity of estrogenic action that is superimposed on the feminizing effects and is responsible for many of the features of the menstrual cycle.

During pregnancy, the placenta becomes an organ of internal secretion, producing estrogens in such excess that quantities are excreted in the urine. Horses and other members of the genus *Equus* produce enormous quantities of estrogens. A pregnant mare will excrete more than 100 milligrams of estrogens daily in her urine. The stallion, despite his evident virility, will deposit more of these "female" hormones in the environment than the mare, or for that matter, more than almost any other animal. The explanation is that both male and female animals produce androgens and estrogens. If either hormone is present in quantities that are not needed, the excess amounts are metabolized, decomposed, or excreted.

The estrogens are steroids differing only slightly in their chemical structure from the male steroids. Estradiol, one of the three female steroids, is the most potent secretion of the ovaries. Another ovarian hormone, progesterone, is produced in abundance during pregnancy. In human females, there is a 100-fold increase in progesterone, produced mainly by the placenta during the later stages of pregnancy. Products that have progesteronelike effects are called progestins—of interest in birth control pills. How progestins, often in combination with a small amount of estradiol, operate to prevent pregnancy can be found in many books and pamphlets on the subject.

Estrus (coming in heat) in animals is brought about by the ovarian hormones acting on the nervous system. Sexual receptivity can be induced almost any time during the estrous cycle by administering estrogens.

Although estrogens have a predominant influence on female sexuality and behavior, they do not have the overpowering effect of androgens. When male rats are treated with estrogen, they retain nearly all of their male behavior. This explains why fairly high levels of estrogen can be present in males without causing an appreciable amount of feminization. Feminization is more readily produced by a lack of androgen than by an overabundance of estrogen. However, a large excess of estrogen is capable of having disrupting effects. This was not thought to be a particular problem until an incident occurred in 1977 in a pharmaceutical factory in Puerto Rico making birth control pills; both men and women developed abnormal sex disorders. Dr. Malcolm Harrington, an epidemiologist with the National Centers for Disease Control, reported that the excess estrogen in the men caused them to have enlarged breasts and reduced sex drive. About half the women in the plant who were tested suffered from intermittent vaginal bleeding. Harrington declared that in large, uncontrolled doses, estrogen "is more potent than we've realized."

A similar incident occurred in 1977 when men working for a suburban Chicago manufacturer of DES (diethylstilbestrol) developed enlarged breasts. Nine of the company's 17 male workers were affected. One of them had his breasts removed by radical mastectomy, a surgical procedure normally reserved for women suffering from breast cancer. Two of the men had become impotent. DES is a potent synthetic female hormone having some use in medicine: it was used primarily as an additive to livestock feed for fattening cattle. It was later identified as a cancer-causing agent. At one time it was administered to women to prevent miscarriage, but its use was stopped when it was found that daughters of many of the treated women developed cancer as long as 20 years later.

Curiously, substances having estrogenic effects are found abundantly in the environment, from flowering plants to the mud in the Red Sea. Some plants contain such high concentrations that they cause serious disorders in animals that feed on them. Sheep in Australia that are allowed to feed on a forage plant called subterranean clover, *Trifolium subterraneum*, become infertile, have abnormal estrous cycles, abortions, or other birth difficulties. Several plants in different families contain the same estrogenic chemical—*genistein*, a weak estrogen, however, compared to estradiol. Another estrogen, *cou-*

mestrol, is found in alfalfa and ladino clover, forage crops for livestock that are also used extensively as feed for dairy cows. Coumestrol is about 30 times more estrogenic than genistein, but still is only one thousandth as potent as estradiol.

Many human food plants are known to have mild estrogenic properties, including some vegetables, cereals, and edible oils. Estrogens in pollen may account for the presumed estrogenic property of honey that is supposed to make it valuable in skin lotions. A widely used flavoring in beverages and medicines, called sarsaparilla, is an extract of the roots of several species of *Smilax*, a member of the lily family. Smilax was an ancient remedy prescribed for a number of ailments including "flagging male virility." The discovery that smilax contains a steroid of a type called *sapogenin* makes it of interest to know whether this relative of the sex steroids has, in fact, testosteronelike activity. But plant estrogens are generally weak compared to human steroid hormones. There is little likelihood that the normal consumption of food in a balanced diet will cause adverse hormone effects in humans.

Male Hormone Domination

We know that the sex of a baby is determined by whether it receives two X chromosomes (XX), in which case it will be female, or an X and a Y chromosome (XY), in which case it will be male. But the early fetus is neither male nor female. It is basically bisexual, or neutral. At first the tissues destined to form the reproductive tracts of the male and female embryos are alike in both sexes, containing the primordial tissues for both male and female reproductive systems. The male gonads have a crucial role in what happens subsequently to the fetus. This has been demonstrated in experiments on animals.

If a male fetus is castrated, it will not develop as a male but will develop exactly in the same way as a normal female. In the female embryo, however, the ovaries are completely passive—the female fetus will develop the same way whether her ovaries are present or not. On the other hand, the fetal testicle produces a substance that inhibits development of the female structures and actively stimulates the development of male structures. Thus the dominant role of the male hormone is evident, even in the early stages of fetal life. This is further demonstrated by the fact that males are less affected than females by treatment with hormones of the opposite sex. When male rats are treated with estrogens, they show little change in behavior, and the same is true of castrated rats when administered estrogens. Therefore, when feminization appears in males, it is not produced as much by the presence of estrogens as by the absence of androgens. However, in females, sexual behavior is closely related to the amount of female hormone present.

Hormones that are present during the fetal period influence the animal for life. When male hormone is administered to a guinea pig fetus by injecting it through the mother's pelvic opening, it acts almost at once to masculinize the external genitalia of the unborn guinea pig. If the fetus is a genetic female, the young guinea pig is masculinized to the point that its behavior after birth is changed to resemble in many ways that of a male. The hormone environment during embryonic, fetal, and early infant life has far-reaching effects on the libido and sexual behavior of the adult animal. If a male rat is castrated at birth, he will show female mating behavior when administered estrogen and will respond very poorly as a male when administered androgen. It is evident that the early secretion of testosterone "imprints" the male response on the developing animal's nerve-endocrine system that causes it to react to testosterone later in life with typical male behavior.

Imprinting also appears in females. If a young girl has an excessive secretion of androgen, she may develop masculine tendencies and have late onset of menstruation. The effect depends on the age at which hypersecretion develops. It may cause masculinization, early semipuberty, or false hermaphroditism. On the other hand, adrenal tumors may have a feminizing effect from excess estrogen secretion. The administration of androgens, or even weakly anabolic agents, in amounts that increase the growth of hair and deepen the voice, will cause irreversible changes in women. When such materials are used in medicine, physicians must guard against giving excessive doses.

The dominating influence of male secretions is seen when twin calves of opposite sexes are born. The female twin is often deformed, having an improperly developed reproductive system and with some secondary male characteristics, whereas the male twin invariably has normal sexual development. The sterile, masculinized female twin is called a *freemartin*. For centuries, farmers and cattlemen were baffled by this event. An understanding of the cause was obtained in 1916 when F. R. Lillie, a scientist at the University of Chicago, studied the uteri of pregnant cattle at slaughter houses. Lillie found that the embryonic twin's placental membranes were fused, so that they shared a common circulation of blood. The development of embryonic testicles is normally in advance of ovary development, so maleness becomes evident at an earlier stage than femaleness. When twins are present, the female fetus is bathed in blood containing a male substance that blocks development of the ovary and causes partial sex reversal. The condition occurs in cows, sheep, goats, and pigs.

The question follows, "What is the male substance that causes the partial sex reversal of freemartins?" It was at first thought that the substance must be the male's androgen, because sex reversal in fish or amphibian larvae can be produced by treating females with androgens or males with estrogens, giving rise to completely fertile sex-reversed individuals. But in mammals, no one has ever been able to cause an ovary to change into a testicle or a testicle to change into an ovary by androgen or estrogen treatment. The fetus has a protective mechanism against the mother's sex hormones that presumably pass through the placenta and circulate in the fetus's blood supply. The male substance responsible for the freemartin effect remains a mystery.

Controlling the sex of an expected child or animal has long been a much desired goal. Control of sex by injection of hormones has been obtained in newly hatched chicks and newborn marsupials, at which stage the animals have both ovaries and testes, neither of them sufficiently developed to determine sex. But normally, the genes for inherited sexuality will determine the hormone balance, which in turn, determines whether the ovaries or the testes will develop. The remarkable inflexibility of the gonads is seen in a type of hermaphroditism common in sows, in which one gonad is an ovary and the other a testicle. Strangely, the sow may be fertile regardless of the size of the male gonad.

Hormones and Homosexuality

The cause of homosexuality is debatable. We have seen that the sex hormones profoundly affect sexual behavior as well as being responsible for anatomical features of the male and female. Testosterone from the testicles stimulates sexual desire in the male, initiates the sexual response at puberty, maintains libido in the adult, and in some species of animals, stimulates male activity during the breeding season. And estradiol brings the female "in heat"—the seasonal estrus.

A widely held view is that homosexual behavior is psychological or of social origin. But some scientists have long thought that there may be a physiological basis for homosexuality and that it is related to the influence of hormones on physiology and behavior. There is evidence that hormones are actually involved, but the extent to which they play a role has not been demonstrated.

When Alan Fisher at the University of Wisconsin injected a soluble form of testosterone into the brain of a laboratory rat so that the hormone came in contact with a specific locus in the hypothalamus, it induced maternal behavior such as nest building, surprisingly regardless of whether the animal was male or female. But when the testosterone was injected into another area slightly to one side of the first, it induced the animal to engage in a male pattern of sexual activity such as mounting, again regardless of whether the animal was male or female. When the hormone was placed between these two sites, the animal became completely confused, unable to decide whether to mount another rat or to build a nest. Clearly, the brain circuits for either male or female behavior are present, and the direction it takes depends on where the hormone exerts its influence.

Several attempts have been made to show a relationship between hormone balance and homosexual behavior. The results were, for the most part, inconclusive and in some cases contradictory. The strongest link between hormones and homosexuality came from work by scientists at the Reproductive Biology Research Foundation at St. Louis, Missouri. They found that the level of testosterone in the blood of men who were exclusively or predominantly homosexual was only 40% of that found in men who were predominantly heterosexual. They also found that most of the homosexuals had below normal sperm counts. The study did not determine, nevertheless, whether the low hormone level was the cause or the result of homosexual behavior, or whether low testosterone is merely incidental to some other physiological condition.

Brian Gladue of the State University of New York, Stony Brook, and co-workers, administered an estrogen preparation to heterosexual men and women and homosexual men. The subsequent secretion of luteinizing hormone (LH) by the homosexuals was intermediate between that of heterosexual men and that of women (estrogen is known to enhance LH more in women than in men). The investigators also found that testosterone was depressed for a longer period in the homosexual men than in the heterosexual men.

The environment can affect hormone action. One example is the effect of crowding on laboratory animals. In experiments conducted by John Christian, when laboratory mice were crowded together in dense populations, the mice were stunted, and reproductive functions were suppressed. In another study, the stress of crowding caused an increase in the size of the adrenal cortex and a decrease in testicular activity. These changes were attributed to suppression of the output of gonadotropin and an increase in adrenocorticotropin by the anterior pituitary. In what is suggestive of a natural mechanism to prevent overpopulation, sexual development is delayed, and the animals may not reach maturity at all if crowding is severe.

When Norway rats were crowded together, a hierarchy developed among the males. Some of them became aggressively dominant and established a territorial domain. Next on the status ladder were those who did not distinguish between appropriate and inappropriate reproductive partners. They would attempt to mount anything that came along: males, females not in estrus, and juveniles. Beneath these were those that withdrew and became completely passive, ignoring and ignored by both males and females. Still other rats became hyperactive and hypersexual, making advances to females without the customary courtship ritual. Some of these eventually became cannibalistic.

A direct cause-and-effect relationship between hormones and homosexuality is not generally accepted, probably because a consistent pattern has not been demonstrated. Treatment with hormones apparently does not change the direction of sexual attraction, although it may change the intensity of sexual desire. Androgen administered to a male homosexual does not change his disposition toward males and leaves his sex urge either unchanged or intensified. High estrogen levels can occur in both males and females because of physiological anomalies. In such cases, the high estrogen in males does not induce homosexuality, nor do high estrogen levels in females induce lesbianism.

However, hormones early in life may have more influence on later sexual attitudes than previously thought. When pregnant rats are subjected to stress, they are apt to give birth to feminized males—males that assume a typically female posture for copulation. One theory for this is that stress causes a change in the normal ratio of androgen from the two sources, gonads and adrenals. Usually the output of androgen is greater from the gonads, but stress during the critical time of sex differentiation in the embryo may reduce or reverse the ratio. In rats the effect is irreversible.

June Reinisch, a psychologist at Rutgers University, found that medicinal hormone treatments to maintain high-risk pregnancies affected the personalities of the children. She studied families in which pregnancies had been treated with synthetic progestin, estrogen, or a combination of the two. Mothers treated with progestin had children who were more independent, sensitive, individualist, self-assured, and self-sufficient than their siblings. Children from pregnancies treated with estrogen were more group oriented and group dependent than their siblings from untreated pregnancies.

Simon LeVay, a neurobiologist at the Salk Institute in San Diego, reported in 1991 that the size of a part of the hypothalamus called the third interstitial nucleus was correlated with male homosexuality. The nucleus studied was only half as large in women and homosexual men as in heterosexual men. This was taken as evidence that people are born gay or straight and not, as some people believe, that being gay is a willful perversity. LeVay's work was questioned on three counts: the size of this particular part of the hypothalamus is difficult to determine because there is no clearly defined border, so the determination is highly subjective; therefore, according to critics, LeVay, who is gay himself, could have been biased (bias is a recognized problem in medical research, necessitating the standard "double blind" study in which neither the patient nor the doctor knows whether the treatment is a drug or a placebo). Furthermore, said the critics, the samples were taken from homosexual patients who had died from acquired immunodeficiency syndrome (AIDS), which left the question of whether the disease itself could have altered the hypothalamus.

In 1993, geneticist Dean Hamer and his colleagues at the National Cancer Institute reported that they had found evidence of a homosexuality gene. They studied 76 homosexual men and found that 13.5% of their brothers were homosexual, compared to about 2% in the general population. They found more gay relatives on the maternal side than on the paternal side, suggesting that in at least some male homosexuals, the trait is passed through the female members of the family, and therefore probably on the X chromosome. A genetic analysis of 40 pairs of homosexuals revealed a region on the X chromosome shared by 33 of the 40 pairs of brothers. That region is where the gene appears to be located. Seven sets of brothers did not share that region of the chromosome, which led Hamer to conclude that homosexuality may arise from more than one cause, possibly environmental as well as genetic.

We have seen that being male or female is determined by the X and Y chromosomes, which direct embryonic cells to differentiate into testicles or ovaries, and these, in turn, determine maleness and femaleness through the action of their secretions of androgens and estrogens. But males and females are not absolute opposites as commonly supposed. Although androgens predominate in males and estrogens predominate in females, both androgens and estrogens are present in both sexes, with resulting degrees of masculinity and femininity. There is no problem as long as the hormones are within normal limits for males and females. Unneeded hormone is excreted in the urine or used for some metabolic purpose. But as shown experimentally, a large excess of either androgen or estrogen can have profound effects. The effect is most pronounced on the fetus, often resulting in drastic changes in sex organs, development, and behavior lasting for life. Thus masculinity and femininity are variable qualities determined by the concentrations of androgens and estrogens to which the organism is exposed in the embryonic and fetal stages as well as during development and adult life.

Chapter 15

The Other Sex Organs

We commonly think of the testicles and ovaries as *the* sex glands. They are the workhorses of the sexual system, the manufacturers of the final products, the front line troops. But they have a relationship with other glands in the body that is more than cooperative; it is one of dependency. The adrenals, the thyroid, the pituitary, and the hypothalamus play a central role in sexuality and are essential to growth, sexual development, and performance.

The adrenal cortex is the gland most closely related to the testicles and ovaries. The adrenals are a pair of structures located immediately above the kidneys, about one thirtieth the size of the kidneys, hardly as large as a person's thumb and flatter. Each adrenal is actually two glands, an inner *medulla*, from the Latin word meaning *pith* and an outer *cortex*, from the Latin word meaning *bark*. The medulla has its embryonic origin in nerve tissue and retains a close association with the nervous system. The cortex grows from the same kind of tissue as the testicles and ovaries and secretes steroids that supplement and reinforce the steroid secretions of the gonads.

The organs that produce steroid hormones—testicles, ovaries, placenta, and the adrenal cortex—are all parts of a coordinated male and female hormone system. Nearly 50 steroids from the adrenal cortex are known. Many of them are probably inactive in themselves but serve as precursors of active hormones. The adrenal glands of both males and females produce both androgens and estrogens. The adrenals secrete some testosterone, but the principal male hormone from the adrenals is one of the so-called weak androgens, aldosterone, which begins to be produced even in the fetal stage. The estrogen produced by the adrenals is secreted in only very small amounts. However, in men and in women after menopause, some of the estrogen comes from conversion of androgens in tissues outside the glands. The thyroid gland has several metabolic functions in addition to those related to sex. When the thyroid first appears in the embryo, it has a duct opening into the mouth, suggesting that its ancestral function was to pour its secretions into the alimentary canal. The duct disappears as the embryo develops, leaving the thyroid a gland of internal secretion. It develops as two lobes in the region of the neck. Its hormone, *thyroxine*, regulates, among other things, growth and early sexual development.

An animal that shows the powerful effect of thyroid hormone is the tiger salamander, *Ambystoma tigrinum*, of North and Central America. These amphibians grow about 8 inches long and are commonly found in parts of the eastern United States. Larval forms called *axolotls* breed in the southwestern United States and in high mountain lakes near Mexico City where they are capable of reproducing without ever reaching maturity. The condition in which reproduction takes place in the larval stage is called *neoteny*, from the Greek *neos*, meaning *young*, and *teinein*, *to stretch*, hence an extended immaturity. Axolotls were at first thought to be distinct species because the eastern tiger salamander reaches full maturity after only a short larval period. But if axolotls are transferred to the eastern United States, they promptly lose their gills, develop lungs, and become typical adult tiger salamanders. Experiments showed that the axolotl is deficient in thyroid secretion, apparently from an undetermined environmental cause.

In many animals, reproductive functions depend greatly on thyroid activity, although it varies with species and the age of the animal. In young males, a decrease in thyroid secretion delays sexual development, but an increase to excessively high levels impairs the reproductive function. Most species can reproduce after removal of the thyroid, but their reproductive capacity is usually impaired.

Rats with deficient thyroid hormone have small litters, and the young are apt to die because lactation is impaired, causing the mother to give insufficient milk.

Neither the adrenal cortex nor the thyroid operates on its own. They are under the control of the pituitary gland, a pea-size but mighty glandular structure at the base of the brain. Like the adrenal, the pituitary is actually two glands, one of them the most versatile and influential sex gland in the body. This is the anterior lobe, or anterior pituitary, also called the adenohypophysis. For its size, it is probably the most important piece of tissue in the body. It secretes several hormones that stimulate other glands into activity. Three of them, collectively called *gonadotropic hormones* or *gonadotropins*, are directly involved in regulating the sex organs. A tropic, or trophic hormone, from the Greek *trophikos*, meaning *nourishing*, is one that influences a specific gland or organ.

The pituitary's gonadotropic hormones are the same in both sexes, but differ in their effects in males and females. When the pituitary is removed from an adult male animal, its testicles will no longer produce spermatozoa, nor will they secrete enough androgen to maintain the accessory sex organs. One of the gonadotropic hormones, the follicle-stimulating hormone (FSH), activates the follicles of the ovary to produce ova in women and stimulates the seminiferous tubules of the testicles to produce sperm in men. Another gonadotropin, the luteinizing hormone (LH), stimulates the testicles to secrete testosterone; in women it has several effects including stimulation of the ovaries to produce steroid hormones and causing an increase in ovarian blood flow. LH also influences the action of FSH on follicle development because if LH is absent, the follicles do not reach their full size and are unable to secrete estrogen. The third gonadotropin secreted by the anterior pituitary is prolactin or lactogenic hormone, also called luteotrophin (LTH). It acts in concert with other hormones to produce a variety of effects in female animals and has the widest spectrum of effects of any of the pituitary hormones.

The anterior pituitary produces two other tropic hormones that influence the reproductive system. One of them, thyrotropin or the thyroid-stimulating hormone (TSH), regulates the output of the thyroid. The other, called the adrenocorticotropic hormone (ACTH) or simply corticotropin, stimulates the adrenal cortex into action. ACTH has far-reaching effects. One of them is to stimulate the output of the chief adrenal androgen, aldosterone. Because the pituitary increases its output of ACTH under conditions of stress, several kinds of stressful situations cause an increase in aldosterone: anxiety, physical traumas, surgery, hemorrhage, excessively high potassium or low sodium intake, constriction of the interior vena cava vein in the thorax, secondary effects from congestive heart failure, cirrhosis of the liver, kidney disease, or simply everyday stressful situations.

The anterior pituitary is sensitively responsive to how hard the glands under its supervision are working. It immediately shuts down the output of a gonadotropin when the gland under its control has produced all of its hormone the body needs. When the concentration in the blood surrounding the pituitary reaches the required level, the pituitary turns off the valve for the gonadotropin and does not open it again until the concentration of the target gland's hormone drops and needs stimulating again. This *feedback* serves to preserve a state of equilibrium in the internal environment of the body by preventing any gland from "getting carried away" once it starts producing.

The pituitary has been called, inaccurately, the master gland because of its vast regulatory power over some of the most important glands in the body. But the authority of the pituitary is subservient to a gland that is still more powerful, the hypothalamus, technically a part of the brain, lying just above and behind the pituitary. The hypothalamus transmits instructions to the pituitary by way of secretions called *release factors*. Several release factors carry chemical signals for the pituitary to "throw the switch" and start producing one or more of the tropic hormones that stimulate the gonads, adrenal cortex, or thyroid, as the case may be. If the hypothalamus is wired to an electroencephalograph, it will show an increase in electrical activity in that part of the brain during sexual excitement. The channel is open in both directions. After stimulation of the hypothalamus of a male rat, he can be made to continue sexual activity long after he would tire of it from natural female stimulation.

Thus the hypothalamus sits at the peak of command in the body's general headquarters, giving orders for this or that gland to pour out its secretions when work is needed. But it cannot give its orders directly to the gland concerned. It must go through channels passing its instructions, via chemical signals, to the chief executive officer, the anterior pituitary, which then relays the message, again via chemical signals of its own, to the target gland. The hypothalamus, then, has under its supervision and control various tissues in the far reaches of the body that determine growth, development, and physiological functions including sexual capability and behavior.

But no boss in a complex organization is without guidelines, restraints, or even orders from a higher authority—which leads us to the sex center itself.

The Biggest Sex Organ

If you are thinking below the belt, forget it. The biggest sex organ is the brain. The brain's action does not depend on the amount of grey matter but on its nerve signals and the amount and kind of chemicals it puts out. As part of the action when a stimulating signal is sent to the part of the brain that activates sexual arousal, the brain begins pumping out potent messages.

Courtship and mating in most animals takes place instinctively without previous experience. But in higher animals and humans, sexual behavior is highly susceptible to suggestibility in ways that are often beyond the control of immediate conscious intent. The physical response in a young human male can be so compelling that it may lead to awkward situations in which all attempts to control an erection by conscious effort fail. Penile erection is one of the best known examples of biofeedback, being influenced to a great extent by mental imagery.

Animals can be conditioned to have either a positive or negative response. If a newly weaned male rat is given an electric shock whenever he approaches an adult female rat, he will continue to avoid females even after he grows up. Because of his early painful experience with females, he is never able to have a normal heterosexual relationship. When normal sexual behavior is blocked in this way, one might expect some form of substitute sexual activity, but there is no evidence that the negatively conditioned rat is inclined to engage in homosexual activity.

Although it is well established that hormones exert a powerful influence on behavior, it is also evident, though less obvious, that the endocrine system can be modified by the external environment through perception and interpretation by the central nervous system. When a female pigeon is exposed to the visual stimulation of an adult pigeon in a separate cage, her oviducts respond by increasing in weight. If the male pigeon is castrated first, there is some stimulation, but the oviduct weights do not increase as much.

External stimulation can overcome exhaustion. Laboratory experiments with rats, chickens, and other animals showed, not surprisingly, that they become exhausted from repeated sexual activity. It takes a hamster 5 days to recover from complete sexual exhaustion. But if an animal that has been exhausted by one sexual partner is presented with a new partner, interest is revived, and much of the old vigor is quickly restored. This has come to be called the *Coolidge effect*. The story is told that when President and Mrs. Coolidge were touring a poultry farm accompanied by separate guides, Mrs. Coolidge asked how often the rooster performed his duty and was told, "Several times a day." She said, "Tell that to Mr. Coolidge." When the message was relayed to the president, he asked whether the rooster did it all with the same hen, and was told, "No, each time with a different hen." The president replied, "Tell that to Mrs. Coolidge."

A male cat will copulate to exhaustion but when given a fresh receptive female, he will experience a recovery of sexual function. Some animals can continue sexual activity almost indefinitely with little rest. If a bull is presented with a series of receptive cows, he will continue to mount them for hours, long after his supply of seminal fluid has run dry.

Animals often show striking sexual preference in the selection of a partner. B. J. LeBoeuf found that when male dogs were tethered and females were allowed to roam freely among them, the females were attracted to some males more than to others. The preferred males received more attention, were sought out more frequently, and were allowed to mount sooner than the others. The females rarely bit or barked at the favored males. Charles Darwin noted the significance of sexual preference in relation to natural selection, and discussed it at length in *On the Origin of Species by Means of Natural Selection* (1859) and *The Descent of Man and Selection in Relation to Sex* (1871). He referred to partner preference as "sexual selection" and pointed out that a hornless stag or a spurless cock would have a poor chance of mating and having offspring.

Even in the lowly fruit fly, much of the male's behavior in typical fruit fly courtship originates in his head. Seymour Benzer and colleagues at the California Institute of Technology bred thousands of fruit flies that were part male and part female in different parts of their bodies. Such genetic patterns are called mosaics. They were able to answer the question, "Is the part of the fly that responds to the allure of the female in the head, body, or the genitals?" If a composite fly had a male head, it would show typical male behavior, following females around and vibrating its wings. Studies by others had also pointed to the brain as the source of courting behavior. However, completion of copulation depends on the coordination of control centers in the brain and other parts of the body.

The behavior of gorillas in captivity belies the notion that all animals follow a rigidly programmed reproductive pattern. The lowland gorilla is an anthropoid ape in an advanced position on the evolutionary ladder near that of humans. Ronald Nadler, at Emory University in Atlanta, Georgia, paired 13 gorillas (9 females and 4 males) in 20 different combinations for about 2000 encounters over a 3-year period. Only 7 of the females copulated regularly. The 2 reluctant females would copulate only when paired with certain males. A partner preference was apparent.

The couples varied their technique by engaging in different positions. Among the 7 more sexually active females, 4 used only the "female back to male belly" position, 1 used only the "belly to belly" or so-called missionary position, and 2 of them would use either position. Two of the males used only the back to belly position, whereas the other 2 used both positions. Some of the couples, with humanlike innovation, indulged in various modifications of the two primary positions.

The sexes played different roles during courtship and foreplay. The male was always shy and reserved. He would approach the female and touch her gently with the back of his hand. The female, on the other hand, was more physical and assertive. She would back into the male, often with enough force to back him into the wall. She would rub her genitalia against him by raising and lowering her rear end, all the while uttering a high-pitched, fluttering whimper.

The variety of sexual positions and related behavior of gorillas contrasts sharply with the stereotyped performance of animals below the level of apes. In the higher primates and humans, learning plays an important part in courtship and mating. Unlike that in most of the lower animals, sexual behavior in apes and humans is conditioned by social and psychic influences. Sexuality in apes, as in humans, has become to a great extent "cephalized."

We have said that the command center for much of sexual activity is in the hypothalamus. The hypothalamus is believed to be the place where the sensation of emotion originates. It is closely associated with the limbic system, the region deep within the brain that deals with instinct and emotion. The limbic system in higher vertebrates is very similar to that in lower vertebrates, indicating that is has changed very little during evolution of the vertebrate brain and that our emotional behavior was developed early in evolutionary history. This explains why our innate emotional and sexual behavior is very much the same as that in other animals—sometimes referred to as behaving beastly.

We've seen that the pituitary is a high-echelon link in the chain of command. It relays signals to and from other glands of internal secretion and serves as the executive officer of a vast organization in which thousands of commands are received, interpreted, and transmitted every hour of the day and night to the sex glands: the adrenal cortex, the thyroid, and the gonads. All of these functions are carried on below the conscious level, influenced by the seat of instinct and emotion—the limbic system. But the instinctual behavior is modified in some of the higher primates, especially humans, under conscious control of the brain.

Chapter 16

The Time of Their Lives

Biological clocks regulate much of human and animal behavior. There is evidence that the clock that rings the sexual chimes is a small, obscure gland tucked away deep in the brain.

The estrous cycle in animals, the menstrual cycle in women, and the waves of libido in men and women that ebb and flow with a force equal to the ocean's tides are not voluntary conditions subject to whim or intent. They are largely dependent on the regulatory authority of deeply rooted biological systems beyond conscious control of volition or reason. What governs the rhythm and where the governor is located was a mystery that puzzled people for centuries, but there is now substantial evidence that the timing center for much of sexual physiology—if indeed there is any one center—resides in a small, and in many ways curious, knoblike outgrowth of the brain called the *pineal*. How this tiny piece of tissue could have power over one of the most basic animal compulsions is a story of the most remarkable evolutionary adaptations of an anatomical organ in nature.

Rhythms pervade the world of nature. They are evident in the cyclic pattern of physiological functions and behavioral patterns of almost all animals. Many of the biological rhythms have obvious cause and effect; others have obscure origins that are still unfathomed and resist the best scientific efforts to unravel their mysterious relationships. The mundane routine of sleeping, eating, excretion, work, social activities, recreation, physical exercise, mental or emotional states, mood, and love all show obvious rhythmic patterns.

The cyclic nature of much of human activity lies deeper than physical performance. It extends to the physiology of the cells and the nervous and circulatory systems, even though we may not be conscious of it because we cannot feel many of the changes taking place. Body temperature rises and falls throughout each day by about 1.5° to 2°, and differs in different parts of the body. There is a daily rhythmic fluctuation in the rate of cell division and in the activity of the enzymes in various tissues of the body, especially in the liver where enzymes make crucial biochemical conversions. The chemicals called neurotransmitters, responsible for conduction of nerve impulses from one neuron to the next, rise and fall in rhythmic fashion in the brain and spinal column. The rhythmic nature of several kinds of electrical brain waves can be measured and recorded. Even the effectiveness of many drugs depends on the time of day they are administered.

We live in a 24-hour cycle, in a day that is part light and part dark, with various rhythmic patterns superimposed. The daily performance of sleep and wakefulness, feeding and resting, and related behavior or physiological activity of daily occurrence are part of a biological *circadian* rhythm, so-called from the Latin *circa*, meaning *about*, and *diurnus*, meaning *of a day*, hence *a rhythm of about a day*. The behavior of mammals is related to their breeding cycles. Most mammals mate during definite breeding seasons, which usually come in the winter or spring. The males of many species can copulate any time of year, but the female will mate only during the part of her cycle when she is in estrus or "heat." When she is not in heat, she will not permit the male to make advances. Estrus is accompanied by changes in the female's vagina and uterus and in her secretions. Ovulation usually occurs during estrus or shortly after, but the time varies among species. The number of times a female comes in heat during the year depends on the species. Dogs, foxes, and bats have only a single estrus during the breeding season, whereas estrus occurs repeatedly in field mice and squirrels. A female rat comes in estrus about every 4 days; a female house mouse about every 4 to 6 days; a cow every 21 days;

and a female dog about every 6 months. However, if any of them mate and become pregnant, the cycle is interrupted. Rabbits do not go through the typical estrus cycle because the female can breed at almost any time. The human female is receptive throughout her reproductive cycle, but in the other higher primates, the estrus period determines when she is receptive to the male.

Frequency of breeding also varies with the species. Small rodents such as mice usually have several litters each year. Field mice have been known to have as many as 17 litters of four to nine mice in one season. Carnivores, by contrast, usually have one litter of two to five young during the year. Very large mammals, such as horses, cattle, and elephants, have only a single young at a time, and often births are less frequent than every year.

Light is known to influence hormone systems. There is a daily rhythm in the activity of the adrenal glands and their production of steroid hormones. In many birds and mammals that breed annually, the growth of their gonads and their breeding cycles are brought on by an increase in sunlight in the spring. Trout normally breed in the autumn, but if the day length is artificially increased in the spring and then decreased in the summer to resemble early fall, they will spawn in the summer as much as 4 months early.

The physiology and behavior of many animals are influenced by moonlight. It is common knowledge that several types of human behavior correlate with phases of the moon. The magic of the silvery moon has long been known to deeply influence moods and behavior, including human love life. But a causative relationship between moonlight and the human menstrual cycle has never been demonstrated. Other forces may have come into play in establishing this approximately monthly rhythm in our ancestral primates, or perhaps it has simply been modified and become less rigidly programmed through evolutionary time.

Many biological rhythms in animals and plants are related to obvious physical cycles such as night and day, the phases of the moon, the ebb and flow of the tides, and the annual changes of the seasons.

When physiological and behavioral rhythms are directly dependent on such cyclic features of the environment, the rhythm is said to be *exogenous*, that is, the biological clock is regulated externally. When the biological rhythm seems to be independent of any detectable exogenous physical influence, it is described as *endogenous*, or internally regulated. So the biological clock that regulates a particular rhythm can be exogenous, endogenous, or a combination in which the regulation is partly external and partly internal. The sum total of external and internal oscillators can cause an entire population of animals to behave rhythmically. This can be seen even in some of the higher vertebrates. Wolf packs often move around the perimeter of their hunting territory with such regularity that their arrival at a particular point can be timed to within half an hour.

The endocrine glands are closely attuned to circadian rhythms. In female rats, the amount of prolactin—the lactogenic hormone that promotes milk production—follows the circadian rhythm of their pituitary glands. In humans, secretions of the adrenal cortex are under the rhythmic control of the adrenocorticotropic hormone (ACTH) output of the pituitary gland. A circadian rhythmic surge of testosterone under the influence of the adrenal gonadotropic hormone results in high levels of the hormone in early morning.

Clocks for some of the human rhythms appear to be in the limbic lobes of the cerebral cortex, called the *limbic system* from the Latin *limbus*, meaning *border*, a region deep within the brain that deals with emotional and instinctive behavior. Two tiny groups of neurons, together called the suprachiasmatic nucleus (SCN), located deep in the brain, are known to generate some of the circadian rhythms. Perhaps subject to the dictates of the SCN, at least one of the timers that has regulatory authority over sexual activity is the *pineal* gland. Its name derives from the Latin *pineus*, meaning *of the pine*, and refers to its conelike shape. In humans, the pineal is about 8 mm long, and is located far back in the brain. The pineal is surrounded by a network of large blood vessels, resulting in a high rate of blood flow, second only to the kidneys when compared weight for weight.

The pineal was an object of speculation as early as the third century BC, when Hirophilus thought it might function as a valve to regulate the flow of thought in the brain. Galen, a Greek physician and anatomist of the second century AD. described the pineal in his writings. He theorized that the pineal, which seemed to differ from brain tissue, had a function much like that of lymph nodes. Anatomists continued to puzzle over the pineal for 2000 years. The French philosopher René Descartes had an idea more than 300 years ago that the pineal's function is to interpret visual perception. It was an uncanny shot in the dark that proved to be partly right. There is compelling evidence that the pineal is derived from a third eye of a remote ancestral three-eyed "monster," and that its function is still related to light. Descartes postulated a connection between the pineal and the eyes by means of "strings," a relationship that has proved to be essentially true if you interpret strings as meaning nerve fibers.

The first indication that the pineal might be related to sexuality came in 1898 when a German pediatrician, Otto Heubner, published a report that a young boy who had a tumor of the pineal experienced abnormal early puberty. Reports of many similar cases followed. It was found later that injury to tissues surrounding the pineal could cause precocious puberty by reducing the activity of the gland. But injury to the pineal itself could, instead, stimulate its function and cause delayed sexual development. Despite this apparent contradiction, it became clear that the function of the pineal had something to do with regulating the gonads. Then in 1918, a discovery was made by Nils Holmgren, a Swedish anatomist, who dissected the pineals from frogs and dogfish. He was astonished to find when he examined them under a microscope that they contained cells similar to the light-sensitive cells called cone cells in the retina of the eye and that nerve fibers led from them. Clearly, they were sensory cells having something to do with light sensitivity. Holmgren was the first to suggest that the pineal might be a type of third eye in cold-blooded vertebrates. Later studies by others proved this assumption to be right. Experiments by Eberhardt Dodt and his colleagues in Germany showed that the pineal of the frog is sensitive to certain wavelengths of light and that perception of light by the pineal is transformed into nerve impulses. The pineals of mammals, however, do not contain these light-sensitive cells.

Because the pineal of mammals is devoid of the light-sensitive cells found in cold-blooded vertebrates, it would be reasonable to assume that its function has lost all relationship to the third eye. But additional studies on the pineal of laboratory animals revealed that the mammalian pineal's function is, in fact, related to light. An important clue was turned up by Virginia Fiske at Wellesley College. She reported that when she exposed rats to continuous light for several weeks, it caused a decrease in the weight of the pineals. We will soon see how this fits in with other findings. For example, when female rats were exposed to continuous light for several weeks, the ovaries increased in weight, and there was an increase in the occurrence of estrus.

Humans and other mammals, as well as birds, many fishes, and amphibians, have a single pineal, but in some of the more primitive vertebrates, the structure is not a single unit but a system of two parts. One part, which takes the form of an outgrowth from the main gland, is the most superficial part of the brain, lying just beneath the epidermis at the top of the head. In some frogs and lizards and other cold-blooded vertebrates, this structure takes the form of a small, eyelike organ with a lens and retina. The organ is called a pineal eye, parietal eye, or simply a "third eye."

A highly publicized example of a reptile with a third eye is the New Zealand tuatara, *Sphenodon punctatum*. The tuatara is called a living fossil because it is the only survivor of a group of lizardlike creatures that flourished during the Triassic period and became extinct (except for the tuatara) at least 100 million years ago. The tuatara survives on one or two islands in the Cook straits where it is protected by the New Zealand government. It reaches a length of about 2 feet when full grown and has several primitive characteristics including a pineal eye and a unique skull configuration. Its pineal eye, which shows evidence of a retina, lies below a small hole covered with scales at the top of the skull. A similar hole in the skull is found in many of today's reptiles and amphibians and in fossil fishes. The tuatara is one of the best examples of evolutionary stagnation, showing practically no evolutionary change for the past 140 million years.

The lamprey, *Entosphenus tridentatus*, is another primitive vertebrate that has an elaborate middle pineal eye with a clear lens and pigmented retina, besides a regular pair of lateral eyes. This and other examples give strong evidence that the fishlike relative of the common ancestor of all higher vertebrates including humans was a three-eyed "monster." But in the higher vertebrates, the pineal has evolved into

Adult sea lampreys are eellike animals with sucking mouths, and are parasitic on fish. Lampreys are not much more advanced on the evolutionary scale than the hypothetical organisms believed to be the ancestors of all vertebrates, including humans. Courtesy of Michigan Department of Natural Resources.

The Pacific lamprey, *Lampetra tridentata*. Behind the lamprey's nasal opening, a clear area in the pigment of the skin shows a third, or pineal, eye. Redrawn by Vicki Frazior from Daniel J. Miller and Robert N. Lea, *Guide to the Coastal Marine Fishes of California*, Fish Bulletin 157, California Department of Fish and Game, 1972.

more than a vestigial third eye. It has become a gland of internal secretion with other important functions, including that of a timer.

The first experimental evidence that the pineal secretes a hormone came when two investigators at Johns Hopkins University, Cary McCord and Floyd Allen, extracted the juice from the pineals taken from cattle. They added some of the extract to water in which tadpoles were swimming and found that the tadpole's skin became lighter in color. Later studies by others led to the isolation and identification of the pineal substance that caused the tadpole's skin to blanche. Aaron Lerner and co-workers at the Yale University School of Medicine purified the extracts from 200,000 cattle pineals, and in 1954 after 4 years of work, identified a compound named *melatonin*. Melatonin is now known as a skin-lightening agent in tadpoles, embryonic fish, and some of the other lower vertebrates. It is found in amphibians in the regular paired eyes as well as the pineals. Its action is to cause clumping of the pigment granules inside the pigment cells called melanophores, leaving spaces for light to pass through. Melatonin does not have this action in mammals because they do not have cells containing mobile pigment granules.

The effect of the pineal in mammals became evident from studies made by Richard Wurtman and Julius Axelrod in collaboration with Elizabeth Chu of the National Cancer Institute. They found that when rats were injected with tiny doses of melatonin, starting just before puberty, the effect was the same as if they were injected with the pineal extract. In either case, the estrus cycle was slowed, and the ovaries lost weight. Conversely, the estrous cycle could be accelerated by removing the pineal, and this acceleration could be reversed by injecting melatonin. It became clear that melatonin is a hormone produced by the pineal, secreted into the bloodstream and adversely affecting a distant organ, the ovary, reducing its activity. Just how melatonin did this remained a mystery until the relationship between light and the periodicity of sexual activity became known.

A clue came with the observation that female rats have a daily rhythm in their sexual excitability. When a female rat that is sexually receptive is in the presence of a male, she raises her rear end by arching her back in a characteristic way called lordosis, from the Greek *lordos*, meaning *bent backward*. Lordosis is taken to be a measure of sexual excitability. If a female is given access to a sexually active male each hour of the day—a different male on each occasion— she is most receptive at two times during the daylight hours: shortly after light, and shortly before dark.

This phenomenon was further clarified when it was found that the daily periodicity could be manipulated. Biologists working on the sexual behavior of guinea pigs found it annoying that the female's sexual cycle almost always began in the middle of the night, reaching a peak very early in the morning. The task of tending the experiments was made easier when it was discovered that the cycle could be reversed by keeping the animals in a dark room during the daytime and lighting it at night.

When male hamsters are kept mostly in the dark by restricting them to only 1 hour of light each day for a month, their testicles atrophy. The same effect can be produced by blinding them even in normal daytime light. Roger Hoffmann and Russel Reiter at the Edgewood Arsenal in Maryland showed that, in either case, testicle deterioration can be prevented by removing the pineal. When rats are exposed to constant light for a day or two, melatonin production falls to as little as one fifth of that produced by animals kept in total darkness. Light can penetrate the skull, which is not totally impervious to it, but the effect is not from light striking the pineal, because when rats are blinded with the head exposed to light, melatonin production does not decline but remains at a high level. Thus the effect is only indirectly on the pineal, from light entering through the eyes.

But the natural environment of most mammals, including rats and humans, is not one of constant light or constant darkness. This left the question, "Will the normal cycle of light and darkness affect the rhythmic function of the pineal?" To find out, Richard Wurtman and Julius Axelrod put their rats in a cage where the lights were on from 7 AM to 7 PM and in the dark all night. They found that by midnight the melatonin production had doubled or tripled. This did not occur when the lights were left on until midnight.

The human output of melatonin also fits a rhythmic pattern. Canadian investigators A. C. Greiner and S. C. Chan found that the melatonin concentration began to increase in the evening and continued through the night, and then started to fall in the morning, reaching its lowest around noon. This is similar to a 24-hour rhythm found in rats, quail, and chickens. The human output of melatonin is not much more than that of rats, even though the human pineal is 100 times larger.

Melatonin does not just appear in the pineal spontaneously. Its production is dependent on the action of two enzymes whose function is to synthesize the hormone. Therefore, the low production of melatonin during continuous light is due to the inhibiting effect of light (acting through the pineal by way of the eyes) on a melatonin-forming enzyme. This, in turn, was found by David Klein and Joan Weller of the National Institute of Child Health and Human Development to be regulated by another neurotransmitter, norepinephrine. Melatonin is synthesized from serotonin, an important brain neurotransmitter, which is present in the pineal in higher concentrations during the day than at night.

In the mammalian embryo, the pineal forms as an outgrowth of the brain, but by the time it reaches full development, its only connection is to the sympathetic branch of the autonomic nervous system, the part of the nervous system that is ordinarily not under voluntary control and is highly sensitive to changes in both the external and internal environments. Light striking the eye stimulates the retina, and the information is transmitted to the pineal along nerve pathways. If the sympathetic nerve connection with the pineal is cut, the normal effect on estrus and ovaries does not occur.

Summarizing the evidence, we have seen that melatonin is a skin-lightening agent in frogs, tadpoles, and some other lower vertebrates but not in mammals. And we have seen that when melatonin is administered to rats, it causes changes opposite to those produced by either removal of the pineal or exposure to light. When rats—male or female—are exposed to an increase in light, there is a decrease in pineal weight and melatonin output, as well as changes in several other functions. In the female rat, the action of the pineal's melatonin (produced mainly at night) is to inhibit the growth of the ovary, and retard the occurrence of estrus. An increase in light, which reduces the output of melatonin, causes an increase in sexual activity. In the male, an increase in light permits the testicles to produce at full capacity instead of shrinking and declining when the animal is kept in the dark. Removing the pineal has about the same effect as exposing the animal to increased light, whereas blinding the animal has the opposite effect.

How does this jigsaw puzzle of information fit together to explain the effect of the pineal on human sexuality? Clinical observations in humans are consistent with the findings in laboratory animals. Reduced activity of the human pineal can cause precocious puberty, and excessive stimulation of the pineal can result in delayed sexual development. The explanation is that secretions of the pineal inhibit production of gonadotropin by the anterior pituitary with consequent adverse effects on the functioning of the gonads. The inhibiting role of melatonin on the function of the female gonads suggests that this hormone may determine menstruation in human females as well as the timing of estrus in other animals.

In humans, cyclic peaks in libido of both men and women are well recognized, but the frequency and intensity vary so much from person to person that a clear pattern defies description. In some women, the rise and fall of desire is related to the menstrual cycle, but other cycles are superimposed that may have an overriding influence. Some men have cycles that peak in days, weeks, or months, whereas others may have several peaks a day. In both men and women, the cycle may differ from time to time in the same person. Much depends on the intensity of other interests and the mental and physical effort expended on them.

Superimposed on this individual variability is the effect of night and day, light and darkness, on the pineal's output of melatonin: an increase in the dark suppressing the gonads, and a decrease during light permitting the gonads to function with full power.

Is the moral to this, contrary to popular belief, that too much night life is bad? Not necessarily. Men and women are partially emancipated from the animal stereotypes that hold reproductive functions in a programmed pattern. When the biological clock that regulates the sex drive runs down, the swing of the pendulum can sometimes be reinvigorated by a change of scenery, a vacation, a new partner, or simply relaxing at the shore or in the woods. Stimulation that reaches the brain through the senses, or originates in the brain itself, can rewind the biological clock that plays the sexual carillon.

We see in the pineal the evolution of an organ from a third eye to a gland of internal secretion but one that did not emancipate itself completely from dependence on light. The primordial third eye may have been used primarily for vision, but as lateral eyes increasingly took over the function of seeing, the third eye became increasingly a gland of secretion until in mammals, no trace remains of the eye's cone cells, and the effect of light is only indirect—through the lateral eyes by a sympathetic nerve connection. The origin of the pineal's sexual function remains a mystery.

Part IV

Beyond Nature

Chapter 17

Surrogate Sires

The breeding and consequent evolution of domestic animals has advanced a long way since Charles Darwin made his studies of animal breeding. Great progress has been possible by a more efficient way than simply mating males and females—namely, artificial insemination. The method has become both a science and an art in the breeding of valuable livestock. The practice is also well established and socially acceptable for women who cannot conveniently become pregnant any other way, or who for various reasons prefer the method.

A celebrated case was that of Kim Casali, a New Zealand-born cartoonist, who was expecting a baby 17 months after her husband, Roberto, had died of cancer. Her pregnancy was made possible by artificial insemination from a deep-freeze sperm bank. Kim, who created the syndicated cartoon series "Love Is," had fallen in love with Roberto Casali when she was working as a waitress in Los Angeles. After marrying, they settled in England where they had two of the three children they wanted. Then tragedy struck. Roberto had incurable cancer. Before he died, he contributed to a sperm bank. Kim, at age 35, said, "The baby will be his last gift to me." The birth announcement forthrightly told that the baby's parents were, "Kim and the late Roberto (posthumously by artificial insemination)."

Pregnancy while the husband is away is sometimes accomplished under happier circumstances. Dr. Edward Tyler of Los Angeles reported the results from artificial inseminations: 68 children born of mothers impregnated with frozen sperm stored up to 2 years. One of the fathers was a soldier in Vietnam who wanted his wife to become pregnant while he was away.

Most artificial inseminations are done with fresh sperm to impregnate women whose husbands for one reason or another are unable to supply either viable sperm or sperm in sufficient quantity. Except when there is good reason to have a close relative furnish the sperm, the donor is selected by the doctor, presumably after thorough screening for genetic background and medical history. Often the donor is a medical intern. In all such cases, the identity of the participants is the most sacrosanct of secrets. The identity of the donor remains unknown to the wife and husband, and they, in turn, remain unknown to the donor. But there is danger in the method of selecting donors. It is an open secret that some donors provide multiple samples of sperm. One donor claimed that he had made about 300 sperm donations between 1976 and 1984. Paul, who asked that his last name not be used because sperm-bank babies looking for their natural fathers might try to find him, was retired as a donor at the age of 35. The sperm bank thought at that age he was over the reproductive hill. A well-supervised program will limit the number of sperm samples a donor can provide, about 24 donations or enough for three pregnancies. Even then there is always a chance that male and female offspring from the same donor will meet, resulting in marriage between half-brother and half-sister. Paul thought that there should be a national registry of sperm donors similar to the registries in Australia, Sweden, New Zealand, and England, where donors are required to be made known when necessary. A medical scandal erupted when it was discovered that a California physician with a large practice had used his own sperm for nearly all his artificial inseminations. Because most of his patients were residents of nearby neighborhoods, the chance of some of the children eventually becoming romantically involved was very great. The doctor protested that he had done nothing illegal, but most people thought it was highly unethical.

The pioneering experiment in artificial insemination was done in

1785 by an Italian scientist-priest, Lazzaro Spallanzani, when he successfully inseminated a female dog with dog semen. He had previously established that semen is necessary for fertilization, a fact not recognized by many people at the time. He included his observations and results of his studies as part of a large two-volume book on physiology published in 1780. Charles Bonnet, a wealthy French-Swiss lawyer whose hobby was the study of natural history, had rediscovered parthenogenesis in aphids and had done a lot of speculating about the generation of offspring. Intrigued and apparently alarmed by Spallanzani's experiment, Bonnet wrote to the abbé, "What you have discovered may some day be applied to the human species . . . and it will not be taken lightly."

Indeed, artificial insemination is not taken lightly. It has become a useful procedure for having children when the usual method of impregnation by coition is impossible or impractical for various reasons. Traditional concepts of fatherhood and the natural concern about having children sired by an unknown man introduce an emotional element, so it is understandable that many people have qualms about using artificial insemination. But there are no such reservations about using the method for livestock breeding. Artificial insemination of farm animals and other valuable livestock, even pets, has developed into a science.

The Canine Cryobank and Animal Fertility Clinic in Escondido, California, does a thriving business in a world grossly overpopulated with dogs and cats of every description. But the Canine Cryobank has a high-class clientele. It banked sperm from sled dogs of Susan Butcher, four-time winner of Alaska's Iditarod race, and Bill Cosby's schnauser, Fat Albert. Chevy Chase brought his golden retriever to the clinic for artificial insemination. The Cryobank was proud of Sarah Lee, named for being the first champion Corgi conceived by sperm from the freezer.

It is estimated that at least three fourths of the nation's dairy cows are bred by artificial insemination, and there is a large and thriving business in exporting semen. The storage of frozen bull semen has even led to crime. Thefts are common, sperm rustlers sometimes getting away with a tankful of sperm worth as much as $100,000 or more.

A bull can father more calves by artificial insemination because his semen can be used to impregnate more cows. Fertility equal to that from natural mating can be obtained with fewer spermatozoa. Several hundred cows can be impregnated with the semen from a single ejaculate from a bull, and if necessary, the semen can be put in cold storage for future use. Insemination with semen is more practical and more economical than hauling around a valuable bull worth thousands of dollars and turning him loose in a pen of heifers.

Semen can be collected by using a surrogate female called a "dummy" or "teaser" and an artificial vagina. Though any resemblance to a real-life cow would require some imagination, it makes no difference to the bull once he gets the hang of it. A more recent, and now more common, method of getting the semen is to stimulate the bull to ejaculate with an electroejaculator consisting of a rectal probe fitted with electrodes. This method, which includes several modifications, is effective with bulls and rams. The artificial vagina is constructed with a liner for holding warm water because chilling would injure the sperm. There is also provision in the artificial vagina for controlling the pressure. Much of the success of artificial insemination of farm animals is made possibly by nature's fondness for reproductive redundancy. A normal male ejaculation puts out many times the number of spermatozoa needed to impregnate a female. A single ejaculate of a good bull may contain as many as 15 billion sperm, enough for more than 1000 cows. (In humans, a normal ejaculate will contain from 100 million to 500 million sperm). Of the domestic animals, the boar puts out the largest number of sperm per ejaculation—30 to 60 billion—which can add up to 120 to 150 billion spermatozoa per week. The boar's semen is also the largest in volume—150 to 200 ml (about three fourths of a cup). Of the larger animals, stallions produce a larger volume of semen that do bulls, but the sperm concentration is less, resulting in the number of spermatozoa per ejaculate being about the same.

In addition to the volume of semen and the number of spermatozoa, there are striking differences in the size and shape of the penis. The glans penis of the boar is shaped like a corkscrew to accommodate the sow's cervix, which is long with twisting ridges and is constricted during estrus. Thus the semen is sprayed in all directions during coitus. To accomplish artificial insemination, a catheter with a corkscrew tip is used, and it is rotated until it fits into the ridges. The sow requires a large volume of semen containing about 2 billion spermatozoa, so one ejaculate from a boar can inseminate only 20 or 30 sows.

Stock breeders have known for a long time that there are great differences among male animals in their ability to make a female pregnant. Failure can be costly because animals come in estrus only periodically, and an entire breeding cycle is lost if the male fails to produce. Scientists at the University of California demonstrated that poor quality semen is responsible for many of the failures. They obtained semen from 465 bulls, and among the young ones—2 years old or younger—about 1 in 12 had semen with a reproductive efficiency of no more than 30%. In older bulls—3 to 6 years old—poor quality semen was much more common. In deficient bulls, the sperm had low motility, there were quantities of abnormal sperm, and there were variable percentages of dead sperm. Most of the abnormal sperm consisted of normal heads detached from their tails, but there were all kinds of abnormal sperm, including sperm with malformed heads. Three of 53 older bulls had one testicle atrophied and no longer functioning.

The presence of leukocytes in semen from some of the bulls suggested that an infection was the probable cause of the problem. However, unknown environmental factors, possibly with cumulative effects, may have been the cause. These bulls had only one tenth the number of sperm found in normal bulls. The investigators found that there was apt to be a conception rate of no more than 30% if fewer than 43% of the sperm were alive, if fewer than 30% were motile, or if more than 35% were abnormal.

Spermatozoa age rapidly under normal conditions. When samples are taken from bull semen and stored at 4°C (39°F), their motility decreases exponentially with time. One half of the spermatozoa will cease to function in less than 24 hours. But when bull or boar semen is to be used fresh, if properly prepared and stored at a little above freezing, it can be used for at least 3 days. The sperm of most animals is killed by freezing unless treated properly for storage, but oddly enough, human sperm is exceptionally hardy at low temperatures. Scientists working with human semen were surprised when they found that a few untreated spermatozoa were still alive when the semen was thawed after deep freezing at extremely low temperatures. A breakthrough that greatly extended the usefulness of artificial insemination was the finding in 1949 that glycerin would afford the spermatozoa protection against freezing and thawing. This made it possible to store semen almost indefinitely. The method used for livestock breeding almost worldwide is to put the semen in glass or plastic ampules placed in liquid nitrogen which holds the temperature at −196°C (−321°F). Samples of semen from bulls long dead are stored this way in a semen bank at the Cattle Breeding Center of Cambridge University.

Artificial insemination of birds seems like a flight of fancy, but it has been going on since the 1940s. Two falconry hobbyists at Greenacres, Washington, developed the method to a fine art. Robert Young, a physician, and Lester Boyd, a Washington State University zoology technician, liked to engage in falconry—one of the most ancient sports—the training and use of falcons as birds of prey to bag other animals. But environmental restrictions made it illegal to trap and domesticate falcons from the wild. Building up a stable of falcons by breeding them in captivity is not easy, either, because the birds are finicky lovers, engaging in a complicated courtship ritual, including food exchange. Fortunately for bird breeders, many kinds of birds are susceptible to "imprinting" and can be manipulated to look upon humans as sex objects. Young and Boyd, taking advantage of this idiosyncrasy, performed the mating ritual for the reluctant male falcons until they were in a mood for romance. To do that, the human "mate" bows, postures, warbles seductively, and offers food. If the male falcon falls for the deception, he deposits his semen in a specially constructed container that seems to suit his fancy. The sperm must be used within 12 to 24 hours, so Young and Boyd had it rushed to Greenacre by carrier pigeon, the cheapest, fastest, and most reliable transportation available for the 160-mile round trip. Females were impregnated using a plastic tube.

A great deal of research has been done on reproduction in mink. Mink have a vicious temper and must be reared in individual cages to prevent their injuring one another. There would be great savings if the animals could be reared in colonies, like chickens or rabbits. Scientists thought they might be able to use tranquilizers or other methods to calm them down, but everything they tried failed.

Richard Aulerich at Michigan State University wondered if it would be possible to breed calmness into minks, either by selective breeding or by crossing a mink with a closely related animal that is less vicious by nature. Crossing one species of animal with another has been done and is successful with some farm animals. Crossing horses with donkeys is about as far apart as one can go and still get progeny. Even then, the offspring is almost always sterile, although there have been rare reports of mules—the hybrid progeny of a jackass and a mare—becoming pregnant and having foals.

Aulerich thought of the ferret as one possibility for crossing with mink. Ferrets and minks are closely related carnivores, some of them belonging to the same genus, *Mustela*. Aulerich was quoted as saying, "If the offspring has a mink's fine fur and the ferret's disposition we'll call it a 'merret.' If it has the mink's disposition and the ferret's coat, we'll call it a 'fink.'" The cross was tried, but nothing happened.

Although husbandrymen have continued to improve the breeds of domestic animals by the kind of selective breeding practiced for centuries, there is no similar control over the quality of human offspring, although managed eugenics is well within our technical ability. Voluntary eugenics of a haphazard sort is practiced daily, or more commonly nightly, by the natural process of sexual selection.

But the thought of having the parentage of our children dictated by a Commissar of Reproduction or by a committee appointed by Congress is somewhat frightening. Hitler tried it by arranged matings in an effort to produce a strain of super Aryans but without any beneficial results that were noticeable. More recently an effort was made in California to create a sperm bank of Nobel Prize winners and other intellectuals, to be made available to selected distinguished ladies. The sperm bank was founded by California business tycoon Robert K. Graham, a developer of plastic lenses for eyeglasses. Graham was a friend of geneticist Hermann Muller, who was awarded the 1946 Nobel Prize in Medicine and Physiology for his pioneering work on mutations. Muller had advocated the establishment of sperm banks so the genetic endowment of gifted men could be widely spread. After Muller died in 1967, Graham established the Hermann J. Muller Repository for Germinal Choice, a subterranean vault at his 10-acre estate near Escondido, California. Graham had written Nobel laureates asking for donations, and five of them agreed. The only one who made public his donation was William B. Shockley, who shared the 1956 Nobel Prize in Physics and obtained notoriety and condemnation for his contention that blacks are genetically inferior. Most Nobel laureates took a dim or indifferent view of Graham's project. Howard Termin, who won the 1975 Nobel Prize for his work in genetics, was quoted as saying, "Who would want a world full of such people? . . . genetics may have nothing to do with it, so why bother?" Linus Pauling who won the 1954 Nobel Prize in Chemistry and another Nobel Prize for Peace in 1962, opined, "The old-fashioned way seems still the best to me." Women were given a choice based on characteristics and history, without the names of the donors being divulged. According to a news story February 29, 1980, three women had been successfully inseminated and more than a dozen other women had expressed an interest.

Managed eugenics in a democracy would involve problems of gargantuan proportions. In view of the performance of government agencies in many of the other less intimate endeavors in which the technical, theoretical, and philosophical aspects are relatively simple, few people would want to trust the judgment of a government expert on which of 100,000 or more human genes would be the answer to the brave new world. But dramatic changes in the lives of people worldwide brought about by technology may be causing a dangerous trend in human evolution that will make it prudent to intervene and modify the process of natural selection.

The dramatic success of medicine in prolonging the lives of people who would otherwise die too early to reproduce enables them to pass on physical and mental defects to their progeny. In an earlier time, many of them would have died early in life, leaving no progeny at all. Many diseases are gene-controlled and are transmissible to offspring through the regular channels of inheritance. There is also clear evidence that susceptibility to some diseases is inherited. And, in still other cases, general vitality, probably controlled by a complex of undefined genes, may be a limiting factor in survival.

Contraceptives make it possible to prevent reproduction at will. But for complex sociological reasons, birth control is generally practiced more widely and more stringently by societies that have demonstrated the highest qualifications for survival, longevity, and success. Thus the two greatest contributions of the biological sciences: germ control and sperm control, are on a collision course that puts the future of the human species in jeopardy of a decline in hereditary characteristics beneficial for survival.

Homo sapiens may get along well for hundreds or thousands of years with a high genetic load of defective genes. But in times of unusual stress, such as a devastating worldwide war or similar major catastrophe, vital medicines and vaccines may not be available. Many people would die for lack of medical attention. Newborn infants would be hit the hardest, because many of them would lack the inherited immunity that in earlier times would have weeded out the susceptible part of the population. Scarcity of food may wipe out other groups who for generations thrived on abundance, but who may not have the genetic capacity to survive semistarvation, nutritional deficiencies, and associated diseases.

Despite the emotional problems and practical limitations of eugenic control, the management of reproduction by artificial insemination based on voluntary participation may, in the distant future, be the best hope for the long-range welfare of mankind. If that were to take place, it would be the most dramatic evolutionary development in the history of the human species.

Chapter 18

Out of the Womb

"And this," said the Director, "is the Fertilizing Room." It was the Director of Hatcheries and Conditioning speaking in Aldous Huxley's novel *Brave New World*. The DHC had opened a door to reveal racks upon racks of test tubes. "This week's supply of ova. . . ," he explained, as he opened another door.

When Aldous Huxley wrote his 1931 novel about how people would relate to the technology of the future, his fictional world seemed like a fanciful pipe dream. But scientific truth has a way of overtaking fiction, and what was intended as an entertaining view of the future of mankind has turned out to be partly prophetic. The moral, ethical, and humanitarian problems have become more real than fictional. Let us have a look at what is going on in the laboratories of scientists in the not-so-brave new world.

It is not surprising that progress in transplantation of embryos was made in animal experiments and livestock breeding earlier than in humans. A goal was to improve productivity of beef cattle by implanting embryos from a donor to a recipient cow. The method is used commercially to increase the number of offspring from genetically superior animals, but the cost limits its use to exceptionally valuable animals. It occurred to L. E. A. Rowson, at the Agricultural Research Council's station near Cambridge, England, that a high-quality female would be able to produce more offspring than if she had to go through the normal gestation period of several months before being able to conceive again. Instead, her ova could be transplanted into other females as often as produced. In 1962, Rowson and his colleagues transplanted fertilized ova from two Border Leicester ewes into the oviduct of a rabbit. The rabbits were flown to South Africa where the eggs were again transplanted back into sheep, this time into two South African Dorper ewes, which gave birth to healthy Border Leicester lambs. The technique was called *artificial involution*.

The economic possibilities in the realm of animal husbandry are far reaching, especially since the discovery that mammalian embryos can be stored by freezing them. A cow can ordinarily give birth to only one calf per year, perhaps six or seven during her lifetime. The reproductive capacity of a cow could be greatly increased if she did not have to carry her unborn calf through the gestation period of 281 days. By foisting the burden on a healthy cow less valuable genetically, the "real" mother could produce far more offspring than she would be able to do on her own. Frozen embryos from a valuable breeding cow would be implanted in less valuable "incubator" cows to carry the fetus through the gestation period, leaving the real mother cow free to produce a never-ending supply of embryos.

The idea was put into practice by the British Agricultural Research Council's Unit of Reproductive Physiology and Biochemistry near Cambridge. In June 1973, a bull calf named Frosty was born from a deep-frozen embryo that had been implanted into a foster mother cow. The midwife group, under the direction of Ian Wilmut and L. E. Rowson, had frozen 21 embryos at $-196°C$, then transplanted them, after thawing, into 11 foster mothers. Only one of the embryos, Frosty, survived to term, illustrating the difficulty of producing "test tube" offspring on a production-line scale. Companies specializing in the procedure offer the service for a fee, which may come to several thousand dollars for each calf. Heifers derived from sound stock of good breeds will bring $100,000 or more when sold for breeding purposes.

The most successful technique is to "superovulate" the donor animal with hormone treatment to stimulate follicle development and release of eggs by the ovary, yielding as many as 30 eggs instead of one. The eggs are fertilized *in situ*, that is, within the donor's reproductive tract, by artificial insemination, following which the embryos are removed surgically for transplanting into a foster mother. Implantation can occur only if the foster mother is in estrus.

One cow named Fernhame Ned Oceana, Ned for short, gained fame by producing 50 calves in 4 years, many of them with price tags in five figures. The 14-year-old Holstein-Friesian was able to produce 33,000 pounds of high-butterfat milk each year. The success of Ned and other high-performance cows attracted the attention of tax accountants, who persuaded the Internal Revenue Service to accept cows as "farm machinery" that could be depreciated, resulting in large tax savings to investors.

A snag to more extensive use of the methodology is that no one has been able to fertilize bovine eggs *in vitro*.* If that could be accomplished, it would greatly extend the usefulness of transplantation. Eggs and sperm from valuable livestock could be used to produce embryos in abundance, to be frozen for shipment and transplantation into foster mothers anywhere in the world.

One purpose of embryo transfers, mostly experimental, is to increase twinning. Gary Anderson and co-workers at the University of California in Davis started transfering embryos in 1973. The procedure is to remove an embryo from a cow surgically and transfer it into the uterus of another cow. If an embryo is implanted in the uterus of a cow that is already pregnant, she will have twins. One of the problems in making embryo transplanting commercially feasible is to develop an embryo culture system that will allow more than a few hours between donor and recipient. A freezing technique similar to that used for sperm is one possibility. Live births of mice have been obtained from embryos frozen for several days. The biggest problem with cattle is that a female twin is usually sterile if the other twin is a male (see freemartin). Therefore, a method of determining the sex of

---

*The culture of an organism's cells or tissues separate from the body, such as in a test tube, is called *in vitro* culture, from the Latin *vitrum*, meaning "glass."

the microscopic embryo before it is transferred would be crucial. A nonsurgical method of removing the embryo would make embryo transfers more widely applicable.

Meanwhile, the successful demonstration of embryo transplantation in livestock was not lost on scientists working in human reproduction. Fertilization of human eggs outside the womb has important medical benefits. When sterility is caused by failure of the ova to pass down the fallopian tube, oocytes can be removed from the ovary and transplanted back into the uterus after fertilizing them *in vitro*. It is estimated that nearly 1.5 million women in the United States alone have the type of infertility caused by blocked fallopian tubes.

Scientists saw early in the game that in order to get ova in the right stage of maturation for fertilization outside the body, it would be desirable to grow them in laboratory culture. An encouraging step was made in 1935 when Gregory Pincus of Harvard University, later at Worcester Foundation for Experimental Biology, and his co-workers, found that when they removed oocytes from the follicles of rabbit ovaries, the oocytes would continue to develop in a standard laboratory culture medium. Later, Robert Edwards, a geneticist-embryologist at Cambridge University, using essentially the technique developed by Pincus, succeeded in getting human oocytes (egg precursors) to mature in culture.

The first attempts to fertilize mammalian eggs outside the body were made by removing sperm from the female's genital tract several hours after mating and adding the sperm to eggs previously removed and held for maturation in a suitable culture medium. The first people to do this successfully were two French scientists, Charles Tibault and Louis Dauzier at Jouy-en-Josas, who used rabbit ova and sperm, and published their results in 1954. Then in 1959 M. C. Chang, an American at the Worcester Foundation for Experimental Biology at Shrewsbury, Massachusetts, transferred embryos obtained by *in vitro* fertilization to the oviducts of foster mothers and obtained healthy young rabbits from them. He found that embryos developing from eggs fertilized *in vitro* were perfectly normal as long as the eggs were allowed to complete their maturation in the rabbit. Maturation refers to the development of an oocyte to a mature egg. The human oocyte does not mature until after ovulation, and as in most mammals, is still in the process of maturation when penetrated by a sperm. Halving of the chromosome number takes place during maturation.

Meanwhile, work continued on developing suitable conditions for maturation of the eggs *in vitro*, because this would be the easiest way to obtain cultured embryos. One of the more successful experiments was conducted by Yu-Chih Hsu at the Johns Hopkins University School of Hygiene and Public Health in Baltimore, Maryland. Hsu was able to culture mouse embryos from the blastocyst stage to a state of development equivalent to about one half the normal 21-day gestation period. The blastocyst stage is when the embryo is a hollow sphere of cells enclosing a fluid-filled center. Projecting from the inside is a clump of cells that will form the fetus. The embryo is in the blastocyst stage when it becomes implanted on the wall of the uterus. Most of Hsu's work was with embryos grown from eggs fertilized by normal mating of male and female mice, followed by washing them out of the female for culturing. But he was also able to get similar results from eggs cultured *in vitro*. If an ovum is taken from the female reproductive tract when it is mature, *in vitro* maturation can be bypassed.

A human embryo from an egg obtained in this way will develop in laboratory culture to the blastocyst stage in about 6 days. For the embryo to develop further, the blastocyst will have to be transferred to the uterus of the mother-to-be. A problem is to get the fertilized egg at an early stage of embryonic development to attach itself to the wall of the uterus, a step called implantation. For implantation to take place, the embryo's development must be synchronized with the changes that are taking place in the lining of the uterus, ordinarily controlled by the concentration of circulating hormones in the mother-to-be. Procedures to do this were developed in laboratory animals, and fortunately, what is true of mice is sometimes true of men and women. Techniques developed by using laboratory animals can often be adapted to humans with only minor changes. Fertilization of human eggs outside the womb was refined to a fairly standard procedure.

In July 1978, British scientists announced that a young woman was expected to deliver a "test tube" baby by cesarean section. Robert Edwards, a geneticist-embryologist working at the National Institute for Medical Research in London and later at Cambridge University, had been working for several years with cultured embryos. He teamed up with Patrick Steptoe, a gynecologist at the Oldham General Hospital in the industrial northwest of England, to try implanting human eggs fertilized *in vitro*. The experiment was questioned in some circles. But when the baby was delivered, a 5-pound, 12-ounce girl named Louise, on July 25, 1978, people realized that she was born into fame—the first test tube baby. The parents, who lived in Bristol, were Lesley Brown, 30, and her husband John Brown, 38, a truck driver.

The woman who is to become the donor is first given a hormone treatment to induce maturation of the ovarian follicles, and mature oocytes are taken from her ovaries by a surgical procedure called *laparoscopy*. A laparoscope, from the Greek *lapara*, meaning *flank*, is a thin telescope with a built-in light source. When the instrument is inserted into the abdomen through a small incision just below the navel, it gives a clear view of the ovaries and nearby organs. A hypodermic needle is inserted into the largest follicle, and its contents are removed. The operation does not require a general anesthetic but the use of one makes both the patient and the physician more comfortable. The oocytes so obtained are mixed in a laboratory vessel with ejaculated spermatozoa. Ejaculated human spermatozoa are prepared by separating most of the seminal fluid and placing them in saline solution in which the concentration of dissolved salts and other constituents is favorable for culturing. When oocytes are added, the spermatozoa can be seen penetrating the eggs within a few hours.

Another useful application of fertilization outside the womb is to help couples have children even though the husband's output of sperm is below par. The usual method is to freeze and store the sperm from several ejaculations, and then use the pooled sperm for artificial insemination. That procedure would not be necessary with fertilization *in vitro* because it takes only a few spermatozoa to fertilize eggs in culture compared to the large number needed when fertilization is dependent on sexual intercourse.

An attractive application of transplantation from the medical point of view, as well as that of the parents, is to take an egg from a woman who is unable to carry a fetus through the full gestation period and transplant it into a healthy woman for completion of fetal development. Sperm from the donor's husband is used to fertilize the egg, thus maintaining the genetic integrity of the couple's offspring while giving the unborn child a healthy environment in a womb until birth. The procedure, if anything, has been too successful. One problem is that the surrogate mother may develop such an emotional attachment to the child that she is reluctant to give it up to the child's biological parents. The legal ramifications are not clear, and there are those who see moral and ethical issues in it as well.

There can be any one of several reasons for using the *in vitro* fertilization technique. Many older women, for various reasons, want to have babies. Mary Shearing was an athletic 53-year-old grandmother who married a younger man. They wanted very much to have children. She tried to conceive naturally but miscarried a few weeks into pregnancy. Then they found that she was too old to meet the criteria of public agencies for an adoption. The problem with most women older than 50 who want to have children, even after menopause, is not the uterus, which is still capable of implantation and producing a healthy baby. The problem is the aging or complete shutdown of the ovaries. Even if eggs are produced, old eggs are more apt to have genetic abnormalities than eggs from young women. An alternative is for eggs donated by a younger woman to be fertilized by sperm from the husband of the mother-to-be. Mary Shearing announced publicly in October 1992 that she was expecting twins by the procedure. Mary and her husband, Don Shearing, said they paid

$10,000 for the procedure and that the hospital was keeping 10 frozen embryos in case something went wrong. Others had taken advantage of the method earlier, and there appears to be an increasing demand. Researchers at a fertility program sponsored by the University of Southern California reported in the *Journal of the American Medical Association* that their studies give assurance that women, regardless of age, can "safely achieve pregnancy long after the ovaries cease to function."

Another reason for using donated eggs and *in vitro* fertilization is to ensure babies are free of genetic defects such as muscular dystrophy, cystic fibrosis, and hemophilia. One technique, called micromanipulation, is to inject the sperm into the egg using a computerized microscope. This enables the operator to pick healthy-looking sperm and avoid fertilizing eggs with abnormal sperm. One of the causes of infertility is that semen of some men contains a high proportion of abnormal sperm that cannot penetrate the egg membrane. Injection of sperm into the egg by micromanipulation makes it possible to have conceptions that otherwise would be impossible. The method is not safe unless combined with an examination to determine that the sperm introduces no genetic defects.

An advantage to *in vitro* fertilization is that oocytes and embryos can be visually inspected. It was long thought that if a method could be devised to detect chromosome abnormalities, it would be a boon to parents to have a choice in the type of offspring they would have. This was made possible by embryo biopsy, a technique developed by Jaime Grifo and his colleagues at Cornell's New York Center for Reproductive Medicine and Infertility. The method, also called preimplantation genetic diagnosis, is used in conjunction with *in vitro* fertilization. A single cell is removed from the embryo when in the 8-cell stage and is subjected to genetic analysis, making it possible to determine in advance of implantation in the mother whether a genetic defect is present. Removal of a single cell does not impair the embryo's ability to develop into a normal fetus. Preimplantation screening has the advantage of determining potential problems with such genetic diseases as Tay-Sachs, muscular dystrophy, cystic fibrosis, and a number of other genetic diseases before pregnancy. Previously, couples with a 25% to 50% chance of having a child with a genetic disease had to rely on a test called chorionic villus sampling (CVS), or amniocentesis, done at 9 to 16 weeks gestation. If a genetic defect was detected at this stage of fetal development, the parents were faced with the difficult decision of whether to terminate the pregnancy. But preimplantation screening is more expensive than the less desirable methods of amniocentesis or CVS after pregnancy.

The first pregnancy and birth in the United States after embryo biopsy was announced in 1993. The mother made two attempts at *in vitro* fertilization with embryo biopsy and delivered a healthy girl weighing more than 9 pounds.

Ethical and legal problems associated with fertilization of human eggs outside the womb impinge on some of the deepest and most emotional aspects of life and birth. All conceptions are fraught with danger to the mother and the child, but it was feared that fertilization *in vitro* could introduce special risks of a formidable nature. Genetic aberrations might appear as the result of increased handling and perhaps from exposure to a chemical environment that departs from normal. Means had not yet been developed for prior detection of chromosome abnormalities in eggs, spermatozoa, and blastocysts without killing them. So there was worry about the potential teratogenic* effects of various steps in the technique. However, no such adverse effects have been observed in practice.

There were other objections as well. Removing, handling, and culture of eggs, spermatozoa, and embryos, bring about the death of many of them. To many people, the termination of life, including embryonic and fetal life, is an unnecessary and inhumane act of aggression. It has special meaning to those whose religious views make it a moral issue. To some of those who base their views on

---

*Teratogenic, from the Greed *tera*, meaning "monster," and *generatus*, "to generate," refers to something capable of causing congenital malformations.

religious grounds, even the sacrifice of sperm is objectionable. There was also a fear among a few scientists as well as others that the technique could be developed to such a high state of efficiency that it could be used in a program of eugenics by a tyrant or government agency. The power could be used to breed races of supposedly superior leaders or races of human beings designed for specialized tasks and used at the whim of dictators or government.

One proposal for applied eugenics that seems humane enough is that, by mating sperm and egg in a test tube, it might be possible to detect not only genetic traits but also potential defects while the embryo is still in the blastocyst stage. The bad ones could then be rejected, a more humane way than that of the ancient Spartans who left their physically defective male infants to die in a gorge at the base of Mount Taygetus. Taking a cue from Huxley, the obviously healthy and genetically perfect embryos would be cultured under various fetal environments for early conditioning depending on a computerized predetermination of the fosterling's destiny in society.

There are nagging bioethical questions. What are the ethics of using human embryos for research regardless of the stage of development? Do the embryos have legal rights? If so, who would represent them? A panel of scientists? The Bar Association? The Church? The State? The courts have decided that research using fetal tissue is permissible, but such rulings are subject to change. If the Supreme Court decided that uncontrolled research is permissible up to the blastocyst stage, but experiments went beyond the point that the embryo was capable of implantation, would the burden of proof be on the experimenter to show that the embryo's rights were not being infringed?

Finally, if the state of the art develops to the point that a fetus could be brought to term *in vitro*—not very likely—who or what would be its parents? In the bureaucratic laboratories of the brave new world, the genealogic origin of the child would be more than a frivolous question.

Chapter 19

The Evolution Revolution

It is an irony of fate that during evolution some of the most profound changes in the characteristics of organisms, especially microorganisms, are brought on by human intervention—intervention in the sexual function of organisms, if you define sex as the transfer of genetic material from one organism to another in a way that makes the progeny genetically different.

The development of molecular biology and its offspring, genetic engineering, more commonly called biotechnology, is a spectacular demonstration of scientific creativity rivaling, if not surpassing, the invention of the atomic bomb, and with more salutary results. How this achievement came about is a classic of scientific research in which contributions made by many scientists, each building on earlier discoveries, culminated in what is a revolution in the science of biology.

A turning point came when, as mentioned earlier, Oswald Avery, Colin MacLeod, and MacLyn McCarty at the Rockefeller Institute, later Rockefeller University, discovered in 1944 that genes are made of DNA. Nucleic acid was known to be a major constituent of the cell nucleus since 1869 when a Swiss biochemist, Johann Miescher, isolated a substance containing both nitrogen and phosphorus from cell remnants in pus. His teacher at the University of Strasbourg, Ernst Hoppe-Seyler, in disbelief, took up the study. Because the substance seemed to come from the cell nucleus, they called it nuclein, later called nucleic acid. In 1909, a Russian-American chemist, Phoebus

Levene, at the Rockefeller Institute, found that some nucleic acids contained the sugar ribose. In 1929 he found that another sugar, deoxyribose (ribose minus an oxygen atom), was present in other nucleic acids. This led to recognition of two groups of nucleic acids: ribonucleic acid (RNA) and deoxyribonucleic acid (DNA). The DNA, with some exceptions, was all in the nucleus, whereas RNA was in the nucleus and in the cytoplasm outside the nucleus. RNA was later found to be present in large amounts in the ribosomes, small organelles scattered through the cytoplasm of the cell.

Martin Kossel, a student of Hoppe-Seyler, took up work on nucleic acids in 1879 and showed that when nucleic acids were broken down, purines and pyrimidines (nitrogen-containing compounds) were among the breakdown products. He identified two purines (adenine and guanine) and three pyrimidines (thymine, cytosine, and uracil). These are now commonly referred to as nitrogen bases. The structure of purines had been worked out in detail in the late 1880s by a German chemist, Emil Fischer. Alexander Todd, a Scottish chemist working at Manchester University in the early 1940s, confirmed and expanded on Levene's conclusions about nucleic acid by synthesizing the nucleotides (units of nucleic acid) and comparing them with naturally occurring nucleic acids. He found them to be identical. Todd was knighted in 1954 and was awarded the 1957 Nobel Prize in chemistry. The stage was set for findings that led directly to the determination of the structure of DNA by James Watson and Francis Crick in 1953.

A crucial discovery was made by Erwin Chargaff, an Austrian-American biochemist at the Columbia University College of Physicians and Surgeons. He set out in the late 1940s to determine the quantity of each of the nitrogen bases in a particular nucleic acid. He analyzed a large number of nucleic acid samples, and by 1951, a pattern had become clear. Generally, the number of adenine units was equal to the number of thymine units, and the number of guanine units was equal to the number of cytosine units. This information was vital to Watson and Crick when they set out to deduce the structure of deoxyribonucleic acid (DNA).

Meanwhile, Maurice Wilkins, a New Zealand-British physicist at King's College, London, undertook the study of DNA fibers by x ray diffraction. Information on structure can be gained by the way x rays are diffracted by the regular spacing of atoms or units in a crystal, making it possible to determine the size of the units, their spacing, and much about the structure of the crystal. Wilkins's associate, Rosalind Franklin, a physical chemist with a specialty in x ray crystallography, concluded that the x ray diffraction pattern of DNA was consistent with a helical form (like a spiral staircase) and that it showed the phosphate groups to be on the outside. Linus Pauling at the California Institute of Technology had obtained evidence in 1951 that molecules of fibrous protein such as collagen were in the shape of a helix, and he postulated that the DNA molecule would be helical also. Franklin gave an x ray diffraction photograph of fibrous DNA to Wilkins who showed it to James Watson, apparently without Franklin's permission. The professional relationship between Franklin and Wilkins was not a happy one, partly because Wilkins apparently thought of her as his assistant, and partly because women were traditionally excluded from some institutional activities at King's College.

Watson, an American biochemist, had gone to Cambridge University hoping for a grant, and ended up in the Cavendish Laboratory under biochemist Max Perutz. Watson was a child prodigy, a radio "whiz kid" who entered the University of Chicago at the age of 15, graduated in 1947, and received his Ph.D. at the University of Indiana in 1950. He had intended to be an ornithologist, but during his graduate studies, his restless mind developed multiple interests, embracing genetics, microbiology, and biochemistry. At Cambridge he met Francis Crick, a physicist who had turned to molecular biology as a graduate student.

Watson and Crick found that they had common interests and almost at once teamed up to speculate on the structure of DNA. They knew from Todd's work that the backbone of nucleic acid consisted of alternating sugar and phosphate units. The x ray diffraction studies of Wilkins and Franklin indicated that the DNA molecule was too wide

(20 angstroms) to be a single strand of nucleic acid (an angstrom is 0.00000001 centimeter). Watson and Crick theorized that the DNA molecule was a double helix and that the nitrogen bases extended inward from each of the sugar-phosphate backbones. The purine and pyrimidine nitrogen bases are of different width, so for the molecule to be a uniform width of 20 angstroms as shown by Franklin's x ray diffraction picture, the bases would have to come together in a particular way. The uniform width could be maintained if an adenine base on one backbone always came opposite a thymine on the other, and a guanine from one backbone always came opposite a cytosine from the other. That arrangement would also fit with Chargaff's finding that the number of adenine and thymine units is equal, and the number of guanine and cytosine units is equal. In this configuration, adenine and thymine would be held together by two hydrogen bonds, and guanine and cytosine would be held together by three hydrogen bonds, thus holding together the two halves of the DNA molecule to form the double helix.

The structure of DNA worked out by Watson and Crick was published in the British journal *Nature* in 1953. Watson, Crick, and Wilkins shared the 1962 Nobel Prize in medicine and physiology. Franklin, whose work was consistently underrated, did not share in the prize, even posthumously. She had died 4 years earlier of cancer at the age of 37.

The Genetic Code

A flurry of activity followed announcement of the Watson-Crick structure for DNA. But questions remained. The one gene, one enzyme rule developed by Beadle and Tatum in the 1940s still appeared to be valid. But how did the nucleotide units in the double helix prescribe which of a thousand or more enzymes should be manufactured? What was the code? Enzymes are proteins formed of one or more chains of amino acids called polypeptides, usually containing hundreds of amino acids. There are only 20 kinds of amino

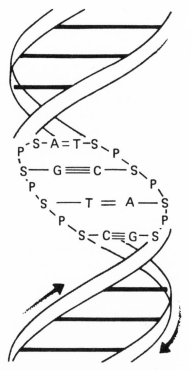

The Watson–Crick model for the DNA double helix, showing the phosphates on the outside and the nitrogen bases on the inside. The two strands are held together by hydrogen bonds, two hydrogen bonds between adenine and thymine, and three hydrogen bonds between guanine and cytosine. Drawing by Vicki Frazior.

acids, but there are millions of possible sequences in which they can be arranged along a chain. Francis Crick theorized that a code of three bases, or a triplet, for each amino acid would satisfy the requirement, and experiments proved this to be correct. Many scientists contributed to cracking the code, and the solution to the complete code came with surprising swiftness.

In 1956, George Palade, a Romanian-American physiologist at

the Rockefeller Institute, showed that numerous small bodies in the cytoplasm called microsomes ("small bodies") consisted largely of RNA, and were therefore renamed *ribosomes*. It was soon recognized that ribosomes were where proteins were manufactured. But an unsolved question remained. If DNA was the repository for genetic information, how did the information get to where the cell could use it to produce protein? François Jacob and his associate, Jacques Monod at the Pasteur Institute suggested that a "messenger RNA" could serve as a template to carry the DNA blueprint from the nucleus to the cytoplasmic ribosomes where proteins are produced. Later experiments by Marshall Nirenberg demonstrated the existence of messenger RNA, usually written mRNA. Then Mahlon Hoagland, a biochemist at the Harvard University Medical School, located small RNA molecules in the cytoplasm, each of which had the ability to combine with one particular amino acid and no other. These were called transfer RNA, or tRNA. Each molecule of tRNA had its own triplet that joined with the complementary triplet on the mRNA, and a protein was built up, amino acid by amino acid.

Severo Ochoa, a Spanish-American biochemist working at New York University College of Medicine, produced a synthetic RNA by incubating nucleotides that would be expected to form RNA with an enzyme that he had extracted from a strain of bacteria. The reaction product had a jellylike consistency as one would expect of a batch of long molecules made of units of nucleic acid strung together. But Ochoa's synthetic RNA differed from natural RNA in an important way. He had begun with only one kind of nucleotide and ended up with an RNA with only one kind of nitrogen base instead of nucleotides containing the four nitrogenous bases in natural RNA: adenine, cytosine, guanine, and uracil.

Marshall Nirenberg, a biochemist at the National Institutes of Health, with the assistance of Heinrich Matthaei, took Ochoa's work a step further. They prepared synthetic RNA containing uracil as the only base, so the nucleotides would read . . . UUUUU . . . indefinitely. The only possible triplet was UUU. When Nirenberg used this as messenger RNA, it produced a product with phenylalanine as the only amino acid. It was clear that UUU was a triplet, or codon, for phenylalanine. Nirenberg had discovered the first entry in the dictionary of the genetic code. Har Gobind Khorana, an Indian-American chemist at the University of Wisconsin, took up work on the code by the same method. He and Nirenberg, working independently, worked out almost the entire genetic code for all of the amino acids. For example, GUA (guanine-uracil-adenine) on messenger RNA is the triplet for the amino acid valine; CAU (cytosine-adenine-uracil) specifies histidine, and so on. By 1966, the genetic code was completely deciphered. Nirenberg and Khorana shared with another chemist, Robert Holley, the 1968 Nobel Prize for Medicine and Physiology.

Each combination of bases in a triplet, called a codon, gives the signal for a specific amino acid to be incorporated into a protein. Most of the amino acids have more than one codon. There are also triplets that are used as start and stop signals. There is more DNA in chromosomes than is needed to code for proteins. These excess nucleotides, constituting probably 95% to 98% of the DNA, have no known function and are called "junk" DNA, but this may be a misnomer because it cannot be assumed that all the junk DNA has no purpose.

Genetic Engineering

Discoveries leading to deciphering the genetic code, and those that followed, empowered humans to change the genetic makeup of organisms faster and in some ways more dramatically than nature, sometimes called "genetic engineering." The potential consequences for good or evil were seen to be so great that some people, including scientists, became alarmed at the ability of scientists to "play god." They feared that scientists, already under suspicion for having brought such monstrous devices as atomic bombs into their lives, might in their enthusiasm for genetic engineering create new forms of life with reckless abandon and release into the environment organisms that would be dangerous and uncontrollable. Despite the perceived risks, researchers pressed on, but with restraints.

A major breakthrough came in 1970 when Hamilton Smith, a microbiologist at Johns Hopkins University, discovered enzymes that acted as submicroscopic chemical scalpels, capable of slicing the DNA molecule at specific sites. Smith had noticed that a bacterium he was working with had a defense mechanism that rapidly broke down foreign phage DNA. A cell-free extract of the bacteria broke down *E. coli* DNA but not DNA of the bacteria from which he obtained the extract. He determined that the enzyme acted at specific cleavage sites. Smith's colleague at Johns Hopkins, Daniel Nathan, produced a map of the enzyme's cleavage sites on the circular DNA from a virus named SV40. By repeating the experiment with other enzymes, he mapped many different cleavage or so-called "restriction" sites. Enzymes responsible for the cleavages were called *restriction enzymes*. Smith and Nathan shared the 1978 Nobel Prize for Physiology and Medicine.

By 1976, other researchers determined the sequence of more than half of the 5,200 base pairs of the DNA of the SV40 virus, using a technique developed by two-time Nobel laureate Frederick Sanger, a biochemist at Cambridge University. A few years later, the entire sequence was determined for the Epstein-Barr virus DNA, 172,282 base pairs long. Improved methods are used to determine the genomes—the total sequence of base pairs, or complete set of genes—of higher organisms. Several research organizations started projects in the late 1980s to work out the human genome, an immense undertaking that coalesced in 1990 into the Human Genome Project (HGP), which soon became international in scope under the leadership of James Watson as the first director. Policy disputes with the head of the administrative agency, the National Institutes of Health (NIH), mainly over patents, led to Watson's resignation about 3 years later, but the project continued. The scope of the task is evident from the estimate that the human genome contains 3 billion nucleotide pairs.

Of greater immediate interest to the general public is the fact that the cleaver enzymes—restriction enzymes—make it possible to obtain known fractions of DNA, along with their genetic information, according to the method of Smith and Nathans, and to put them together in other combinations by means of an enzyme called DNA ligase, to form "recombinant DNA." Paul Berg, a biochemist at Stanford University, pioneered the work on recombinant DNA, for which he was awarded a share of 1980 Nobel Prize in Chemistry. Genetic engineering became a reality in 1973 when Stanley Cohen at Stanford University and Herbert Boyer at the University of California at San Francisco inserted genes from the African clawed toad into bacteria. Cohen and Boyer shared the Nobel prize with Paul Berg for their pioneering experiment. Now, organisms with new genetic characteristics are being created routinely by introducing recombinant DNA into their genetic machinery. The workhorse in this enterprise is our old friend *E. coli*, used because the bacterium is easy to culture and manipulate genetically. But other organisms, yeast for example, are used as well. The use of *E. coli* was at first of great concern because the bacterium finds a congenial home in the human intestine, and some strains are dangerous pathogens. Paul Berg was one of several authorities in the field who advocated that research in recombinant DNA be regulated. An 18-month moratorium was agreed on, during which guidelines and regulatory procedures were put in place. But subsequent work convinced most workers that the risk was exaggerated, so controls were eventually slightly relaxed. For one thing, it turned out that the strain of *E. coli* used in genetic engineering has become so modified that it probably could not survive in the human gut.

A crucial step in the use of recombinant DNA in biotechnology was gene cloning. A clone is an exact duplicate of an organism. Horticulturists clone plants as a standard practice simply by planting cuttings such as, for example, patented roses. Cloning occurs naturally in several ways. When microorganisms reproduce by fission or budding they produce clones, except for possible mutations. If genes in pieces of recombinant DNA are introduced into bacteria, large numbers of clones will be produced in a short while. At the rate of doubling every 20 minutes or so, a single bacterium will produce billions of progeny containing the cloned genes in a day's time. But a more efficient method of selectively obtaining desired genes from the 100,000 or so genes in a mammal was devised in the mid-1980s by Kary Mullis, a biochemist working for Cetus Corporation in Emeryville, California. The method, called the polymerase chain reaction (PCR), so called because it uses DNA polymerase, a DNA repair enzyme discovered by Arthur Kornberg and co-workers in 1955 at Stanford University, makes it possible to bypass the time-consuming cloning procedure by the use of heat and suitable chemical reagents. The method speeded up and greatly expanded the applications of biotechnology. Kary Mullis was awarded a share of the 1993 Nobel prize in Chemistry. Cetus is reported to have sold the PCR patent for $300 million. James Watson and coauthors, in their book, *Recombinant DNA*, described PCR as "a technical revolution in molecular genetics."

Entrepreneurs saw early in the game that because all living organisms recognize the genetic code and slavishly use it in the production line for the manufacture of proteins, it would be possible to create new forms of life and their products according to predetermined design. The exciting prospect of creating an infinite variety of products to improve the human condition stimulated a flurry of activity in biotechnology unparalleled in science or commerce. Several companies were founded, most by academic specialists at universities who were trained and experienced in the field, with financing by venture capital from investors or by stock issues or both.

The attack on human problems was on several fronts, mainly focused on pharmaceuticals, diagnostic methods, identification of genes causing genetic diseases, agriculture, and to a lesser extent, energy. The first pharmaceutical to come from recombinant DNA was insulin, produced in 1978 by scientists at Genentech, a biotechnology company in South San Francisco. Insulin is a life-saving hormone for millions of diabetics. The natural hormone, produced by specialized cells in the pancreas, was isolated in 1921 by Frederick Banting and Charles Best at the University of Toronto. Banting shared the 1923 Nobel Prize in Medicine and Physiology with a University of Toronto professor, John MacLeod, who did nothing on the project except provide laboratory space. Banting was furious about Best's being excluded and considered refusing the award, but ended up giving half of his share of the prize money to Best. After discovery of the hormone, supplies of therapeutic insulin were obtained from the pancreas of slaughtered cattle and pigs. But animal insulin differs slightly from human insulin in the amino acid sequence, causing some people to have an allergic reaction. Today, human insulin from recombinant DNA, produced by our familiar friend *E. coli*, is in abundant supply.

Another hormone important to growth, development, and health is the human growth hormone (HGH), produced by the pituitary gland. An extreme deficiency of the growth hormone in children causes hypopituitary dwarfism, which prevents children from achieving normal stature. Therapy consists of injections with the hormone. But animal hormones do not work in humans. The only source until recently was human cadavers, an unforeseen consequence of which was the fatal infection of a few children with a virus from one of the cadavers. It took the hormone from 50 cadavers to treat a child (at great cost) for 1 year. Now an abundant supply of HGH is produced by recombinant DNA.

One of the shining successes in medicine is the use of vaccines to prevent infectious diseases. Vaccination against smallpox was practiced in a crude way in early times in Asia and the Middle East. Smallpox was a dreadful scourge that during severe epidemics killed one of every three persons who came down with it. Everyone knew that those who had contracted the disease would never get it again, so many brave souls played the dangerous game of modified Russian roulette by inoculating themselves with pus taken from the blisters of those who appeared to have mild cases. There was an old wives' tale in

England that anyone who caught cowpox, a mild disease of cattle, became immune to both cowpox and smallpox. Edward Jenner, a physician, decided that the idea had to be tested, so on May 14, 1796, he took some fluid from a blister on the hand of a milkmaid who had cowpox and injected it into a boy named James Phipps, who came down with cowpox. Two months later, Jenner inoculated the boy with smallpox. The boy remained healthy, and Jenner became a hero, especially after successfully repeating the experiment 2 years later. Of course, he would have been a criminal if either of the experiments had failed.

The Latin word for cowpox is *vaccinia*, from *vacca*, the word for cow. Jenner coined the word vaccination. Louis Pasteur, who a half century later elevated the art of immunization to the level of a science, honored Jenner by calling his own inoculations for other diseases, "vaccinations." Now that the smallpox virus has been eradicated from the earth, except for laboratory samples at two tightly guarded laboratories, there is no need for smallpox vaccinations. But there is still need for improvements in vaccines. Several children die every year from polio vaccinations and immunization for other diseases. Until the advent of recombinant DNA, mainly two types of vaccines were used: *inactivated* vaccines consisting of infectious agents killed by chemical treatment, and *attenuated* vaccines containing strains of live bacteria or viruses that have lost their virulence. Both types are dangerous because they can become contaminated with live infectious agents. The vaccines work by exposing their protein surfaces to B and T lymphocytes (forms of white blood cells) that become conditioned to initiate the production of antibodies if an infection takes place. A recombinant DNA technique is to produce only the protein surfaces that activate the immune system. They are called *subunit vaccines*, an advantage being that there is no risk of infection. The first subunit vaccine successfully produced was for hepatitis B virus (HBV), which causes severe liver damage and sometimes cancer. There is great interest in developing subunit vaccines for protection from other human and animal diseases without the risk of infection from the vaccine itself.

The concept of a "magic bullet" dates from the work of Paul Ehrlich, a German bacteriologist, who announced in 1910 the discovery of a chemical that was effective against the spirochaete that causes syphilis. Ehrlich's search for a magic bullet was an idea that foreshadowed a class of materials developed decades later for use in chemotherapy, a word also coined by Ehrlich. A potentially major development in biotechnology is the use of antibodies designed to destroy tumor cells and disease agents by acting as carriers of drugs or other therapeutic agents to specific target sites. James Watson referred to them as a new kind of magic bullet. Antibodies bind to a single target, and that one only, among millions of available sites in the body. The engineered products, called monoclonal antibodies (MAbs), so called because they are derived from a single clone line, are used for diagnosis of infections and cancers and for identifying tissues to be treated with radiation therapy. Their eventual use as magic bullets in actual treatment of cancer and immune disorders has great promise.

Biotechnology is no less promising in agriculture. Insects are a major problem in agriculture, causing huge losses in food and fiber at great cost to farmers and, therefore, to consumers. Biotechnology can help reduce both the cost and the reliance on poisonous chemicals. A microbial pesticide, *Bacillus thuringiensis* (Bt), developed commercially in the early 1960s, is effective against several species of insects. Bt produces spores containing crystals of a toxic protein that kills larvae when they ingest it. It is harmless to the insects' parasites and has no effect on humans or other vertebrates. Unfortunately, the effectiveness of Bt does not last long on foliage. It occurred to researchers that if the Bt toxin gene could be introduced into the genome of the plant to be protected, the protective effect could be greatly prolonged. Transgenic tobacco and tomato plants containing the Bt toxin gene did, in fact, kill tobacco hornworms, but modifications are necessary to enhance its effectiveness against most insects. Several ideas are being worked on. One approach is to modify the Bt toxin coding sequence. Another scheme is to redesign the toxin protein to make it more effective.

Improving the quality or the production efficiency of plants and

Hornworms, *Protoparce sexta* and *P. quinquemaculata*, enemies of tomato, tobacco, potato, and related plants, succumb to a toxin produced by transgenic plants containing a gene from *Bacillus thuringiensis* (Bt), which produces the toxin naturally in its spores.

animals occurs on a broad front. One objective is to protect plants from virus diseases by developing transgenic plants that express a protective virus coat protein. Viruses cause great losses in plant productivity. Weeds are another threat to every farmer's production and, despite the harrow and the hoe, account for more pesticide use than any other pest problem. Unlike insecticides, which are designed to kill animal life on or among plants, herbicides (weedkillers) must be able to selectively kill plant life growing among other forms of plants that must be spared. The desired level of specificity is not always attainable. An objective of biotechnology is to produce transgenic crop plants that are resistant to herbicides. Several ways of doing it with recombinant DNA have promise. Another objective in the field of agricultural biotechnology, and one having unlimited possibilities for pleasure and profit, is the production of new colors and forms of flowers and other ornamental plants. Horticulturists have made great changes in the characteristics of ornamental plants through the years by carefully selecting desirable varieties and propagating them by cuttings or grafting. Biotechnology has the advantage of not having to wait for natural mutations to provide the horticulturist with varieties to select.

An early success in agricultural biotechnology was the development of recombinant bovine growth hormone (rbGH) produced by *E. coli*. Growth hormone, also called *somatotropin*, has long been known to increase milk production when injected into cows. But the only source was pituitary glands, making it too scarce and costly to use. Recombinant bovine growth hormone (rbGH) increased milk production and required less feed. A question remained about residues of rbGH in the milk and its safety. After several years of evaluating the risks, the Veterinary Medicine Advisory Committee set up by the Food and Drug Administration, reported to the FDA that recombinant BGH presents an "insignificant" and "manageable" risk. Large dairy associations favored FDA approval, but sixty dairy and grocery companies and the major infant formula producers were strongly opposed. The dispute remains to be resolved.

Despite the great benefits already realized and as yet unrealized, biotechnology has its problems. A brief review of accomplishments and prospects does not reveal the complicated and time-consuming steps used in some of the procedures or the costly research required. Profits for most pioneering biotechnology companies were slow in coming. Several were taken under the wings of large, established chemical manufacturers and nurtured by infusions of capital to sustain their research and development. Products did not always come off the line at prices to the user that would seem to be a boon to mankind. When Genentech came on the market in 1987 with an anticlotting drug for heart attack victims called t-PA, a thrombolytic or so-called clot-buster named Activate, it bore a price tag of $2,200 per dose. An older drug called streptokinase, used for the same purpose, cost $200 per dose. A 3-year study completed in early 1993 showed that Activate was 14% more effective than the older drug (6.3% versus 7.2% deaths), although there was some skepticism that the difference would hold in all cases. When the difference in effectiveness is perceived as slight, agonizing decisions clouded with ethical questions have to be made in the treatment of some patients. The day the t-PA report was released, investors pushed the price of Genentech stock up $4.75 a share in heavy trading on the New York Stock Exchange. Clearly, scientists were not the only ones who had great expectations for rewards from biotechnology, but biotechnology is not an easy road to riches. The time it takes from an idea to a product averages 12 years, with the expenditure of large amounts of money for research and clinical trials, often with disappointments as well as triumphs along the way. As late as 1992, only 3 out of 24 biopharmaceutical firms showed a profit for the year, yet the promise of contributions to society was impressive. By 1993, 17 new pharmaceuticals based on biotechnology had been approved by the Food and Drug Administration, and more than 130 were undergoing clinical testing.

The evolution revolution is in its infancy. Most of it cannot be told because it has not yet happened. It is humbling to realize that Shakespeare was speaking to most of us in Hamlet's line, "There are more things in heaven and earth, Horatio, than are dreamt of. . . ."

Epilogue

The transfer of genes from one organism to another by sexual contact is the essence of inheritance and a major player in evolution. Sex has been going on for a long time—at least 2 or 3 billion years if the behavior of present-day microorganisms is an indication of what happened soon after the dawn of life. The practice was retained throughout evolutionary history, and with relatively few exceptions, became the predominant feature of reproduction in the lives of all higher animals, especially vertebrates. Sex, that is, the transfer of genes from male to female and subsequent transfer to progeny, is apparently an irrevocable part of reproduction, inheritance, and evolution of mammals.

We did not know that organisms as simple as bacteria engage in sex until Joshua Lederberg made the discovery in 1946 that *E. coli* bacteria pass their genes from one to another. He found later than even phage viruses transfer genes from one bacterium to another. The finding that sexual activity is rampant in the microbial world puts a new dimension on inheritance and evolution. The transfer of genes among microorganisms causes virulent strains to spread and accounts in large part for the sudden appearance of populations of pathogenic organisms that are resistant to drugs. The extent to which microbial transfer of genes from one class of organisms to another played a role in evolution is unknown, but it may have been more important than realized.

Human intervention in the transfer of genes from microbe to

microbe, and even between microbes, mammals, and plants, adds a third dimension to the sex imperative. Biotechnology—genetic engineering—by accelerating and expanding the exchange of genes among organisms brings about changes so profound and so quick that it can be called the agent of instant evolution.

As the manipulation of recombinant DNA becomes ever more sophisticated and expanded, new forms of organisms, some of them never imagined, will be created. Will precautions be adequate to avoid bad genes mingling with good ones? Will organisms be released carelessly in the environment? Thus far, scientists have been meticulous at observing the precautions. Let us hope they do not lose sight of the lurking danger while focusing on the limitless possibilities of biotechnology.

Bibliography

Chapter 1

Abel, Ernest L., *Ancient Views on the Origin of Life* (Cranbury, NJ: Fairleigh Dickenson University Press; Associated University Presses, Inc., 1973).

Arrhenius, Svante, "The Propagation of Life in Space," *Die Umschau* **7** (1903), trans. Donald Goldsmith; reprinted in Donald Goldsmith, *The Quest for Extraterrestrial Life* (Mill Valley, CA: University Science Books, 1980), pp. 32–33.

Arrhenius, S., *Worlds in the Making* (New York: Harper & Row, 1908).

Barghoorn, E.S., "The Oldest Fossils," *Scientific American* **224** (1971), pp. 30–42.

Bernal, J.D., *The Physical Basis of Life* (London: Routledge and Kegan Paul, 1951).

Bernal, J.D., *The Origin of Life* (Cleveland, NY: The World Publishing Company, 1967), p. 20, quotation from the letters of Charles Darwin.

Cairns-Smith, A.G., *Seven Clues to the Origin of Life: A Scientific Detective Story* (New York: Cambridge University Press, 1985).

Crick, F.H.C., and Orgel, L.E., "Directed Panspermia," *Icarus* **19** (1973) pp. 341–346.

Crick, Francis, *Life Itself: Its Origin and Nature* (New York: Simon & Schuster, 1981).

Darwin, Charles, Quotation from letter to Joseph Hooker, dated Feb. 1, 1871. Reprinted in H. Hartman, *et al.*, eds., *Search for the Universal Ancestors* (Washington, D.C.: National Aeronautics and Space Administration, 1985).

Day, William W., *Genesis on Planet Earth: The Search for Life's Beginning* (New Haven and London: Yale University Press, 1984).

de Duve, Christian, *Blueprint for a Cell: The Nature and Origin of Life* (Carolina Biological Supply Co., 1991), reviewed in *American Scientist* **80** (July–August, 1992) pp. 387–388.

Farley, John, *The Spontaneous Generation Controversy from Descartes to Oparin* (Baltimore: Johns Hopkins University Press, 1977).

Fox, Sidney W., and Horada, Kaoru, "Thermal Copolymerization of Amino Acids to a Product Resembling Protein," *Science* **128** (Nov. 14, 1958), p. 1214.

Fox, Sidney W., and Dose, Klaus, *Molecular Evolution of Life* (San Francisco: Freeman, 1972).

Haldane, J.B.S., "The Origin of Life," *Rationalists Annual* (1929), reprinted in Donald Goldsmith, ed., *The Quest for Extraterrestrial Life* (Mill Valley, CA: University Science Books, 1980), pp. 28–31.

Hartman, H., Lawless, J.G., and Morrison, P., eds., *Search for the Universal Ancestors* (Washington, D.C.: National Aeronautics and Space Administration, 1985).

Hoyle, Fred, and Wickramasinghe, Chandra, *Lifecloud* (London: J.M. Dent, 1978; Harper & Row, 1978).

Hoyle, Fred, and Wickramasinghe, Chandra, *Diseases from Space* (London: J.M. Dent, 1979).

Hoyle, Fred, and Wickramasinghe, N.C., *Evolution from Space: A Theory of Cosmic Creationism* (New York: Simon & Schuster, 1981).

Hoyle, Fred and Wickramasinghe, Chandra, *Space Travelers* (Cardiff, England: University College Cardiff Press, 1981).

Hoyle, Fred, *The Intelligent Universe* (London: Dorling Kindersly, Ltd., 1983; New York: Holt, Rinehart and Winston, 1984).

Laszlo, Pierre, "Chemical Reactions on Clays," *Science* **235** (1987), pp. 1473–1477.

Lewin, Roger, "RNA Catalysis Gives Fresh Perspective on the Origin of Life" *Science* **231** (February 7, 1986), pp. 545–546.

Miller, Stanley L., "A Production of Amino Acids under Possible Primitive Earth Conditions," *Science* **117** (1953) pp. 528–529.

Miller, Stanley L., "Production of Some Organic Compounds under Possible Primitive Earth Conditions," *Journal of the American Chemical Society* **77** (1955), pp. 2351–2361.

Miller, Stanley L., and Orgel, Leslie, E., Origins of Life on Earth (Englewood Cliffs, NJ: Prentice-Hall, 1974).

Morowitz, Harold J., *Beginnings of Cellular Life: Metabolism Recapitulates Biogenesis* (New Haven: Yale University Press, 1992).

Oparin, A.I., *Origin of Life*, (1936), 2nd ed., trans. Sergius Morgulis (New York: Dover Publications, 1965).

Ponnamperuma, Cyril, *The Origins of Life* (New York: Dutton, 1972).

Schopf, J. William, "Microfossils of the Early Archean Apex Chert: New Evidence of the Antiquity of Life," *Science* **260** (1993), pp. 640–646.

Shapiro, Robert, *Origins: A Skeptic's Guide to the Creation of Life on Earth* (New York: Summit Books, 1986).

Tunnicliffe, Verena, "Hydrothermal-Vent Communities of the Deep Sea," *American Scientist* **80** (1992), pp. 336–349.

Urey, H.C., *The Planets* (New Haven: Yale University Press, 1952).

Waldrop, M. Mitchell, "Goodbye to the Warm Little Pond?" *Science* **250** (1990), pp. 1078–1080.

Weinberg, Steven, *The First Three Minutes: A Modern View of the Origin of the Universe* (New York: Basic Books, 1977).

Chapter 2

Bell, Graham, *The Masterpiece of Nature: The Evolution and Genetics of Sexuality* (Berkeley: University of California Press, 1982).

Burton, Robert, *The Mating Game* (Lausanne: Elsevier; New York Crown Publishers, 1976).

Ghiselin, Michael T., *The Economy of Nature and the Evolution of Sex* (Berkeley: University of California Press, 1974).

Jacob, François, "Evolution and Tinkering," *Science* **195** (1977), pp. 1161–1166.

Margulis, Lynn, and Sagan, Dorion, *Origins of Sex: Three Billion Years of Genetic Recombination* (New Haven: Yale University Press, 1986).

Smith, John Maynard, "What Use is Sex?" *Journal of Theoretical Biology* **30** (1971), pp. 319–335.

Smith, John Maynard, *The Evolution of Sex* (Cambridge: Cambridge University Press, 1978).

Mayr, Ernst, *The Growth of Biological Thought: Diversity, Evolution, and Inheritance* (Cambridge, MA: Harvard University Press, 1982).

Michod, Richard E., and Levin, Bruce R., eds., *The Evolution of Sex: An Examination of Current Ideas* (Sunderland, MA: Sinauer, 1988).

Short, R.V., "Sex Determination and Differentiation," in C.R. Austin and R.V. Short, eds., *Reproduction in Mammals: Book 2 Embryonic and Fetal Development* (London: Cambridge University Press, 1972).

Thurber, James, and White, E.B., *Is Sex Necessary?* (New York: Harper & Row, 1929).

Williams, George, C., *Adaptation and Natural Selection* (Princeton: Princeton University Press, 1966).

Williams, George, C., *Sex and Evolution* (Princeton: Princeton University Press, 1975).

Chapter 3

Austin, C.R., *Fertilization* (Englewood Cliffs, NJ: Prentice-Hall, 1965).

Avery, Oswald, T., MacLeod, Colin M., and McCarty, MacLyn, "Induction of Transformation by a Deoxyribonucleic Acid Fraction Isolated from Pneumococcus Type III," *Journal of Experimental Medicine* **79** (February 1, 1944), pp. 137–158.

Bishop, David W., *Spermatozoan Motility* (Washington, D.C.: American Association for the Advancement of Science, 1962).

Bishop, J.A., and Cook, L.M., eds., *Genetic Consequences of Man-Made Change* (New York: Academic Press, 1981).

Bloom, Barry R., and Murray, Christopher J.L., "Tuberculosis: Commentary on a Reemergent Killer," *Science* **257** (August 21, 1992) pp. 1055–1064.

Brock, Thomas D., and Madigan, Michael T., *Biology of Microorganisms*, 5th ed. (Englewood Cliffs, NJ: Prentice-Hall, 1988).

Burton, Robert, *The Mating Game* (Lausanne: Elsevier; New York: Crown Publishers, Inc., 1976).

Butler, J.A.V., *Gene Control in the Living Cell* (New York: Basic Books, Inc., 1968).

Cohen, Michael L., "Epidemiology of Drug Resistance: Implications for a Post-Antimicrobial Era," *Science* **257** (August 21, 1992) pp. 1050–1055.

Dawkins, Richard R., *The Selfish Gene* (New York and Oxford: Oxford University Press, 1976).

Dobzhansky, Theodorus, *Genetics and the Origin of Species* (New York and London: Columbia University Press, 3rd ed., revised, 1951).

Levy, Stuart B., and Miller, Robert V., *Gene Transfer in the Environment* (New York: McGraw-Hill, 1989).

Eisenstein, B.I., Sox, T., Biswas, G., Blackman, E., and Sparling, P.F., "Conjugal Transfer of Gonococcal Penicillinase Plasmid," *Science* **195** (March 11, 1977) pp. 998–1000.

Engemann, Joseph G., and Hegener, Robert, *Invertebrate Zoology* (New York: Macmillan Publishing Co., 1981).

Ferone, Robert, O'Shea, Mary, and Yoeli, Meir, "Altered Dihydrofolate Reductase Associated with Drug-Resistance Transfer Between Rodent Plasmodia," *Science* **167** (Feb. 27, 1970), pp. 1263–1264.

Gibbons, Ann, "Exploring New Strategies to Fight Drug-Resistant Microbes," *Science* **257** (August 21, 1992) pp. 1036–1038.

Griffith, Frederick, "The Significance of Pneumococcal Types," *Journal of Hygiene* **27** (January 1928), pp. 141–144.

Harold, L.C., and Baldwin, R.A., "Ecological Effects of Antibiotics," *FDA Papers* (February 1967), pp. 20–24.

Hurst, L., and Hamilton, W.D., "Cytoplasmic Fusion and the Nature of the Sexes," *Proceedings of the Royal Society (London), Series B 247* (March 1992), p. 189, cited by Alun Anderson, "The Evolution of Sexes," *Science* **257** (July 17, 1992), pp. 324–326.

Hurst, L., "Intra-genomic Conflict as an Evolutionary Force," *Proceedings of the Royal Society (London), Series B 248* (1992), p. 135, cited by Alun Anderson, "The Evolution of Sexes," *Science* **257** (July 17, 1992), pp. 324–326.

Jacobs, William R., Jr., Barletta, Raúl G., Udani, Rupa, Chan, John, Kalkut, Gary, Sosne, Gabriel, Kiesner, Tobias, Sarkis, Gary J., Hatfull, Graham, F., and Bloom, Barry R., "Rapid Assessment of Drug Susceptibilities of *Mycobacterium tuberculosis* by Means of Luciferase Reporter Phages," *Science* **260** (1993), pp. 819–822.

Judson, Horace Freeland, *The Eighth Day of Creation: The Makers of the Revolution in Biology* (New York: Simon & Schuster, 1980).

Krause, Richard M., "The Origin of Plagues: Old and New," *Science* **257** (August 21, 1992), pp. 1073–1078.

Lederberg, Joshua, and Tatum, E.L., "Gene Recombination in *E. coli,*" *Nature* **158** (1946), p. 558.

Levy, Stuart B., *The Antibiotic Paradox: How Miracle Drugs are Destroying the Miracle* (New York: Plenum, 1992).

Levy, Stuart B., and Miller, Robert V., *Gene Transfer in the Environment* (New York: McGraw-Hill, 1989).

Metcalf, C.L., and Flint, W.P., revised by R.L. Metcalf, *Destructive and Useful Insects*, 8th ed. (New York: McGraw-Hill Book Co., 1962).

Neu, Harold C., "The Crisis in Antibiotic Resistance," *Science* **257** (August 21, 1992), pp. 1064–1073.

O'Day, Danton H., and Horgen, Paul A., *Sexual Interactions in Eukaryotic Microbes* (New York: Academic Press, 1981).

Orgel, L.E., and Crick, F.H.C., "Selfish DNA: The Ultimate Parasite," in John Maynard Smith, ed., *Evolution Now* (San Francisco: W.H. Freeman and Company, 1982).

Chapter 4

Ardrey, Robert, *The Hunting Hypothesis* (New York: Bantam Books, 1976).

Austin, C.R., *Fertilization* (Englewood Cliffs, NJ: Prentice-Hall, 1965).

Austin, C.R., "Fertilization," in *Reproduction in Mammals I. Germ Cells and Fertilization*, ed. C.R. Austin and R.V. Short (London: Cambridge University Press, 1972), pp. 103–133.

Baker, T.G., "Primordial Germ Cells," in *Reproduction in Mammals I. Germ Cells and Fertilization*, ed. C.R. Austin and R.V. Short (London: Cambridge University Press, 1972), pp. 1–13.

Bishop, David W., ed., *Spermatozoan Motility*, Publication No. 72 (Washington, D.C.: American Association for the Advancement of Science, 1962).

Holmes, R.L., *Reproduction and Environment* (New York: W.W. Norton, 1968).

Monesi, V., "Spermatogenesis and the Spermatozoa," in *Reproduction in Mammals I. Germ Cells and Fertilization*, ed. C.R. Austin and R.V. Short (London: Cambridge University Press, 1972), pp. 46–84.

Rikmenspoel, Robert, "Biophysical Approaches to the Measurement of

Sperm Motility," in *Spermatozoan Motility*, Publ. No 72. ed. David W. Bishop (Washington, D.C.: American Association for the Advancement of Science, 1962), pp. 31–54.

Rosenfeld, Albert, "Controls on Male Fertility Now Seem Within Our Reach," *Smithsonian* **8** (July 1977), pp. 36–43.

Rothschild, Lord, "Sperm Movement Problems and Observation," in *Spermatozoan Motility*, Publ. No. 72, ed. David W. Bishop (Washington, D.C.: American Association for the Advancement of Science, 1962), pp. 13–29.

Chapter 5

Abele, L.G., and Gilchrist, S., "Homosexual Rape and Sexual Selection in Acanthocephal Worms," *Science* **197** (1977), pp. 81–83.

Anon., "Early Scientists Assemble Atop an Ancient Rift." *Science* **258** (Nov. 13, 1992), pp. 1082–1083.

Dickerson, Mary C., *The Frog Book* (New York: Dover Publications, Inc., 1906).

Engemann, Joseph G., and Hegner, Robert W., *Invertebrate Zoology* (New York: Macmillan Publishing Co., 1981).

Giese, Arthur C., and Pearse, John S., eds., *Reproduction of Marine Invertebrates* (New York: Academic Press, 1974).

Hickman, Cleveland P., *Integrated Principles of Zoology* (St. Louis: C.V. Mosby Company, 1961).

Jacobson, A.L., Fried, C., and Horowitz, S.D., "Planarians and Memory," *Nature* **209** (1966), pp. 599–601.

Koestler, Arthur, *The Case of the Midwife Toad* (New York: Vintage Books, Random House, 1973).

Noble, Gladwyn Kingsley, *The Biology of the Amphibia* (New York: McGraw-Hill; 1931; Dover Publications, 1954).

Perrill, Stephen A., Gerhardt, Carl H., and Daniel, Richard, "Sexual Parasitism in the Green Tree Frog (*Hyla cinerea*)," *Science* **200** (June 9, 1978), pp. 1179–1180.

Storer, Tracey I., and Usinger, Robert L., *General Zoology*, 4th ed. (New York: McGraw-Hill, Inc., 1975).

Tyler, Michael J., *Frogs* (Sydney, London: Collins, 1976).

Chapter 6

Clausen, Lucy W., *Insect Fact and Folklore*, (New York: Collier Books, 1962).

Comstock, John Henry, *An Introduction to Entomology* (Ithaca, NY: Comstock Publishing, 1925).

Condit, Ira J., *The Fig* (Waltham, MA: Chronica Botanica Co., 1947).

Ehrlich, P.R. and Raven, P.H., "Butterflies and Plants," *Scientific American* **216** (1967), pp. 104–113. (reprinted in Eisner and Wilson).

Eisner, T. and Wilson, E.O., The Insects: Readings from *Scientific American*, (San Francisco: W.H. Freeman, 1977).

Essig, E.O., *Insects and Mites of Western North America* (New York: Macmillan, 1958).

Kerr, Richard, A., "A Half-Billion-Year Head Start For Life on Land," *Science* **258** (November 13, 1992), pp. 1082–1083.

Maeterlinck, Maurice, *The Life of the Bee*, trans. Alfred Sutro (New York: Dodd, 1948).

Maho, Yvon Le, "The Emperor Penguin: A Strategy to Live and Breed in the Cold," *American Scientist* **65** (November–December 1977), pp. 680–693.

Marlatt, C.L., *Useful and Injurious Insects and the Grape-vine Phylloxera* (Washington, D.C.: Government Printing Office, published by the Department of State, 1893).

Metcalf, C.L., Flint, W.P., and Metcalf, R.L., *Destructive and Useful Insects* (New York: McGraw-Hill, 1962).

Muzik, Katherine, and Fedak, Michael, "Birds that Refuse to Freeze," *Sea Frontiers* **22** (July-August, 1976), pp. 223–233.

Pinshow, Barry, Fedak, Michael A., and Schmidt-Nielsen, Knut, "Terrestrial Locomotion in Penguins: It Costs More to Waddle," *Science* **195** (February 11, 1977), pp. 592–594.

Ramirez, B.W., "Fig Wasps: Mechanism of Pollen Transfer," *Science* **163** (1969), pp. 580–581.

Roeder, Kenneth D., ed., *Insect Physiology* (New York: John Wiley & Sons, 1953).

Simmons, P, Reed, W.D., and McGregor, E.A., *Fig. Insects in California* (Washington, D.C.: U.S. Department of Agriculture Circular 157, 1931).

Stonehouse, Bernard, ed., *The Biology of Penguins* (Baltimore: University Park Press, 1975).

Størmer, Leif, "Arthropod Invasion of Land During Late Silurian and Devonian Times," *Science* **197** (September 30, 1977), pp. 1362–1634.

Wendt, Herbert, *The Sex Life of the Animals* (New York: Simon & Schuster, 1965).

Wheeler, William Morton, *Ants: Their Structure, Development and Behavior* (New York: Columbia University Press, 1910).

Young, John Z., *The Life of Vertebrates*, 2nd ed. (New York and London: Oxford University Press, 1962).

Chapter 7

Ayers, Neil, "Up-Sizing: Penis Enlargement Operation Herald a New Era for Men," *Men's Fitness* **9**(2), (February 1993), pp. 20, 22.

Barke, James, "Pornography and Bawdry in Literature and Society," in *Robert Burns: The Merry Muses of Caledonia*, ed. James Barke and Sydney Goodsir Smith (New York: Gramercy, 1959), pp. 23–37.

Carayon, J., "Traumatic Insemination and the Paragenital System," in *Monograph of Cimicidae, Thomas Say Foundation* **7**, ed. Robert Usinger (Philadelphia: Entomological Society of America, 1966).

Crews, David, "The Annotated Anole: Studies on the Control of Lizard Reproduction," *American Scientist* **65** (1977), pp. 428–434.

Crews, David, "Hemipenile Preference: Stimulus Control of Male Mounting Behavior in the Lizard *Anolis carolinensis*." *Science* **199** (1978), pp. 195–196.

Eberhard, William G., *Sexual Selection and Animal Genitalia* (Cambridge, MA: Harvard University Press, 1985).

Harcourt, A.H., Harvey, P.H., Larson, S.G., and Short, R.V., "Testes Weight, Body Weight and Breeding System in Primates," in John Maynard Smith, ed., *Evolution Now: A Century After Darwin* (San Francisco: W.H. Freeman, 1982), pp. 227–239.

Husband, Robert W., "The Reproductive Anatomy of Male *Rhynchopolipus rhynchophori* Parasite Mite of the Palm Weevil," *Micron* **10** (1979), pp. 165–166.

Martin, Robert D., and May, Robert M., "Outward Signs of Breeding," in Julian Maynard Smith, ed., *Evolution Now: A Century After Darwin* (San Francisco: W.H. Freeman, 1982).

Milne, Lorus, and Milne, Margery, *The Mating Instinct* (New York: New American Library, 1968).

Morris, Desmond, *The Naked Ape* (New York: McGraw-Hill, 1967).

Nadler, Ronald D., "Sexual Cyclicity in Captive Lowland Gorillas," *Science* **189** (1975), pp. 813–814.

Petren, Kenneth, Bolger, Douglas T., and Case, Ted. J., "Mechanisms in the Competitive Success of an Invading Sexual Gecko over an Asexual Native," *Science* **259** (1993), pp. 354–358.

Schultz, A.H., *Anat. Rec.* **72** (1938), pp. 87–394, cited by A.H. Harcourt, *et al.*, "Testes Weight, Body Weight and Breeding System in Primates, in John Maynard Smith, ed. Evolution Now (San Francisco: W.H. Freeman, 1982).

Waage, J.K., "Dual Function of the Damselfly Penis: Sperm Removal and Transfer," *Science* **203** (1979), pp. 916–918.

Wigglesworth, Vincent B., *The Principle of Insect Physiology*, 6th ed., rev., (London: Methuen, New York: E.P. Dutton, 1965).

Chapter 8

Elephants

Carrington, Richard, *Elephants* (New York: Basic Books, 1959).

Darwin, Charles, *The Expression of the Emotions in Man and Animals* (Chicago: University of Chicago Press, reprinted 1965).

Douglas-Hamilton, Iain and Oria, *Among the Elephants* (New York: The Viking Press, 1976).

Douglas-Hamilton, Oria, "Africa's Elephants: Can They Survive?", *National Geographic* **158** (2) (November, 1980), pp. 568–603.

Short, R.V., "Species Differences," in *Reproduction in Mammals 4. Reproductive Patterns*, ed. C.R. Austin and R.V. Short (Cambridge, England: Cambridge University Press, 1972).

Williams, James Howard, *Elephant Bill* (London: Hart-Davis, 1950).

Whales

Burton, Robert, *The Life and Death of Whales* (New York: Universe Books, 1973).

Cousteau, Jacques-Yves, and Diolé, Philippe, *Whales*, trans. J.F. Bernard (Garden City, NY: Doubleday, 1972).

McVay, Scott, "Stalking the Arctic Whale," *American Scientist* **61** (1973), pp. 24–37.

Melville, Herman, *Moby-Dick; or, the Whale* (New York: Harper & Brothers, 1851).

Myers, Norman, "The Whale Controversy," *American Scientist* **63** (1975), pp. 448–455.

Norris, Kenneth S., *Whales, Dolphins and Porpoises* (Los Angeles: University of California Press, 1966).

Payne, Roger, "Swimming with Patagonia's Right Whales," *National Geographic* **142**(4) (1972), pp. 577–586.

Samaras, W.F., "Reproductive Behavior of the Gray Whales, *Eschrichtius robustus*, in Baja California," *Bulletin Southern California Academy of Science* **73** (2) (1974), pp. 57–64.

Scammon, Charles M., *The Marine Mammals of the Northwestern Coast of North America* (New York: G.P. Putnam's Sons, 1874).

Scoresby, William, Jr., *An Account of the Arctic Regions, with a History and Description of the Northern Whale-Fishery* (Edinburgh: Archibald Constable, 1820).

Slijper, E.J., *Whales*, trans. A.J. Pomerans (New York: Basic Books, 1962).

Small, George L., *The Blue Whale* (New York: Columbia University Press, 1971).

Walker, Theodore J., *Whales Primer* (New York, 1967).

Walker, Theodore J., "The California Gray Whale Comes Back," *National Geographic* **139** (3) (1971), p. 394.

Wilson, T.C. and Behrens, David W., "Concurrent Sexual Behavior in Three Groups of Gray Whales, *Eschrichtius robustus*, During the Northern Migration off the California Coast," *California Fish and Game*, pp. 50–53.

Chapter 9

Anon., *Sea Frontiers* 22(4) (July-August, 1976).

Anon., *Sea Secrets* 18(1) (1974), p. 5.

Anon., *Sea Secrets* 20(2), fourth series (March-April, 1976), p. 3.

Allredge, Alice, "Appendicularians," *Scientific American* **235**(1) (July, 1976), pp. 94–102.

Ardrey, Robert, *African Genesis* (New York: Dell, 1961).

Barfield, Sahton, "Biological Influences on Sex Differences in Behavior," in *Sex Differences*, ed. Michael S. Teitelbaum (Garden City, NY: Doubleday, 1976).

Daugherty, Anita E., *Marine Mammals of California*, 2nd rev. (Sacramento: State of California Fish and Game, 1972).

Hunt, George L., Jr., and Hunt, Molly Warner, "Female-female Pairing in Western Gulls (*Larus occidentalis*) in Southern California," *Science* **196** (1977), pp. 1466–1467.

Le Boeuf, Burnery, "The Resurgent Elephant Seal May Bear the Seed of its Own Demise," *Los Angeles Times* (January 15, 1978).

Lorenz, Konrad, *King Solomon's Ring: New Light on Animal Ways* (New York: Thomas Y. Crowell, 1952; New American Library, 1972).

Reinboth, R., ed., "Intersexuality in the Animal Kingdom," Papers from a Symposium, Mainz, Germany, July, 1974 (New York: Springer Verlag, 1975), reviewed by Richard L. Miller, *Science* **191** (1976), pp. 845–846.

Seymour, George, "Sheepshead the Favorite Quarry of Sport Fishers," *Outdoor California* (September-October, 1974).

Simpson, George Gaylord, *The Meaning of Evolution*, revised edition (New Haven: Yale University Press, 1967; Bantam Books, 1971).

Thomas, Lowell P., "Coral Reefs, Mangroves, and Dutch Gin," *Sea Frontiers* **20**(5) (1974), pp. 285–293.

Young, W.C., ed., *Sex and Internal Secretions*, 3rd ed. (Baltimore: Williams & Wilkins, 1961), cited by John Paul Scott, *Animal Behavior* (Chicago: University of Chicago Press, 1972).

Chapter 10

Aristotle, *Historia Animalium*, trans. R. Cresswell (London: Bell, 1897), cited by E.C. Hertzler, "Airborne Pheromones," *Science* **162** (1968), p. 813.

Asimov, Isaac, *Asimov's Guide to the Bible, Vol. 2: New Testament* (New York: Avon Books, 1969).

Asimov, Isaac, *Biographical Encyclopedia of Science and Technology* (New York: Avon Books, 1976).

Austin, C.R., *Fertilization* (Englewood Cliffs, NJ: Prentice-Hall, 1965).

Austin, C.R., "Fertilization," in A.R. Austin and R.V. Short, eds., *Reproduction in Mammals 1: Germ Cells and Fertilization* (London: Cambridge University Press, 1972).

Cuellar, O., "Animal Parthenogenesis," *Science* **197** (1977), pp. 837–843.

Darlington, C.D., *Genetics and Man* (New York: Schocken Books, 1964). Matthew 1: 18–23.

Metcalf, C.L., and Flint, W.P., *Destructive and Useful Insects*, 4th ed., rev. R.L. Metcalf (New York: McGraw-Hill, 1962).

Milne, Lorus J., and Milne, Margery, *The Biotic World and Man*, 3rd ed. (Englewood Cliffs, NJ: Prentice-Hall, 1965).

Nordenskiold, Erik, *The History of Biology*, new edition (New York: Tudor Publishing, 1935).

Ohno, S., "The Development of Sexual Reproduction," in C.R. Austin and R.V. Short, *Reproduction in Mammals, Book 6: The Evolution of Reproduction* (London: University of Cambridge Press, 1976), pp. 1–31.

Petren, Kenneth, Bolger, Douglas T., and Case, Ted J., "Mechanisms in the Competitive Success of an Invading Sexual Gecko over an Asexual Native," *Science* **259** (1993), pp. 354–358.

Reese, K.M., "Molly and the Girls," *Chemical & Engineering News*, Nov. 13, 1972.

Ross, W.D., Aristotle Selections (New York: Scribner, 1955).

Rattrey Taylor, Gordon, *The Biological Time Bomb* (New York: The World Publishing Company, 1968).

Rorvik, David, *In His Image: The Cloning of a Man* (Philadelphia: J.B. Lippincott, 1978).

Wigglesworth, Vincent B., *The Principles of Insect Physiology*, 6th ed., rev. (London: Methuen, New York: E.P. Dutton, 1965).

Chapter 11

John and Elisabeth Buck, "Synchronous Fireflies," *Scientific American* **234**(5), (May, 1976), pp. 74–85.

Carlson, Albert D., and Copeland, Jonathan, "Behavioral Plasticity in the Flash Communication Systems of Fireflies," *American Scientist* **66** (1978), pp. 340–346.

Harvey, E. Newton, *Bioluminescence* (New York: Academic Press, 1952).

Lall, Abner B., Seliger, Howard H., and Biggley, William H., "Ecology of Colors of Firefly Bioluminescence," *Science* **210** (1980), pp. 560–562.

Lloyd, James E., "Aggressive Mimicry in Photuris Firefly Femme Fatales," *Science* **149** (1965), pp. 653–654.

Lloyd, James E., "Aggressive Mimicry in Photuris Fireflies: Signal Repertoires by Femmes Fatales," *Science* **187** (1975), pp. 452–453.

Lloyd, James E., "Male Photuris Fireflies Mimic Sexual Signals of Their Females' Prey," *Science* **210** (1980), pp. 669–671.

McCosker, John E., "Flashlight Fishes," *Scientific American* **263**(3), (March 1977), pp. 106–114.

McElroy, William D., and Seliger, Howard H., "Biological Luminescence," *Scientific American* **208** (1962), pp. 76–89.

Morrison, Samuel Eliot, *Admiral of the Ocean Sea* (Boston: Little, Brown and Company, 1942).

Storer, Tracy I., and Usinger, Robert L., *General Zoology*, 4th ed. (New York: McGraw-Hill, 1965).

Tsuji, Frederick I., "Light Production in the Luminous Fishes Photoblepharon and Anomalops from the Banda Islands," *Science* **173** (1971), pp. 143–145.

Zahl, Paul A., "Wing-borne Lamps of the Summer Night," *National Geographic*, July 1962, pp. 48–59.

Chapter 12

Albone, Eric S., *Mammalian Semiochemistry: The Investigation of Chemical Signals Between Mammals* (New York: John Wiley & Sons, 1984).

Ballou, C.H., "Effects of Geranium on the Japanese Beetle," *Journal of Economic Entomology* **22** (1929), pp. 289–293.

Beroza, Morton, *Chemicals Controlling Insect Behavior* (New York: Academic Press, 1970).

Butler, C.G. and Calam, D.H., "Pheromones of the Honeybee—The Secretion of the Nasanoff Gland of the Worker," *Journal of Insect Physiology* **19** (1969), p. 237, cited by Morton Beroza, 1970.

Cardé, R.T., "Utilization of Pheromones in the Population Management of Moth Pests," *Environmental Health Perspectives* **14** (1976), pp. 133–144.

Collins, C.W. and Potts, S.F., Attractants for the Flying Gypsy Moths as an Aid in Locating New Infestations. *USDA Technical Bulletin* **336** 1–43, 1932.

Darwin, Charles, *The Descent of Man and Selection in Relation to Sex*, 2nd ed. (New York: A.L. Burt, 1874).

Dodson, Calaway H., Dressler, Robert L., Hills, Harold G., Adams, Ralph M., and Williams Norris H., "Biologically Active Compounds in Orchid Fragrances," *Science* **164** (1969), pp. 1243–1249.

Evans, William F., *Communication in the Animal World* (New York: Thomas Y. Crowell, 1968).

Fabre, J.H., *Souvenirs Entomologiques*, 10 volumes, 1879–1908 (Paris: Librarie Delgrave).

Fabre, J.H., *The Life of the Caterpillar* (New York: Dodd, Mead, 1916).

Forbush, E.H. and Fernald, C.H., *The Gypsy Moth, Porthetria dispar* (Linn.) (Boston: Wright and Potter, 1986).

Gary, N.E., "Chemical Mating Attractants in the Queen Honeybee," *Science* **136** (1962), pp. 773–774.

Gibbons, Boyd, "The Intimate Sense of Smell," *National Geographic* **170**(3), (September, 1986), pp. 324–360.

Gilbert, Lawrence E., "Postmating Female Odor in *Helioconius* Butterflies: A Male-contributed Antiaphrodisiac?" *Science* **193** (1976), pp. 419–420.

Gough, Paul, "Natural Enemies Used to Fight Insect Ravages," in *The Yearbook of Agriculture* (1975), p. 101.

Hara, Toshiaki J., Ueda, Kazuo, and Gorbman, Aubrey, "Electroencephalographic Studies of Homing Salmon," *Science* **149** (1965), pp. 884–885.

Haworth, A., *Insect Miscellanies* (Boston: Lilly and Wait and Carter and Hendee, 1832), cited by R.T. Cardé, 1976.

Izard, M.K., "Pheromones and Reproduction in Domestic Animals," in John G. Vandenbergh, ed., *Pheromones and Reproduction in Mammals* (New York: Academic Press, 1983), pp. 253–285.

Jacobson, Martin, Beroza, Morton, and Jones, W.A., "Isolation, Identification, and Synthesis of the Sex Attractant of the Gypsy Moth," *Science* **132** (1960), p. 1011.

Jacobson, Martin, "Recent Progress in the Chemistry of Insect Sex Attractants," in *New Approaches to Pest Control and Eradication, Advances in Chemistry Series* **41** (1963), pp. 1–10.

Jacobson, Martin, *Insect Sex Attractants* (New York: Interscience, 1965).

Jacobson, Martin, *Insect Sex Pheromones* (New York: Academic Press, 1972).

Jacobson, Martin, and Beroza, Morton, "Chemical Insect Attractants," *Science* **140** (1963), pp. 1367–1373.

Jacobson, Martin, and Beroza, Morton, "Insect Attractants," *Scientific American* **211**(2) (1964), p. 20–27.

Jacobson, Martin, Beroza, Morton, and Yamamoto, Robert, "Isolation and Identification of the Sex Attractant of the American Cockroach," *Science* **139** (1963), pp. 48–49.

Johnson, Robert E., and Zahorik, Donna M., "Taste Aversions to Sexual Attractants," *Science* **189** (1975), pp. 893–894.

Karlson, P., and Butenandt, A., "Pheromones (ectohormones) in Insects," *Annual Review of Entomology* **4** (1959), p. 39.

Kochert, Gary, "Sexual Pheromones in *Volvox* Development," in O'Day, Danton H. and Horgen, Paul A., eds., *Sexual Interactions in Eukaryotic Microbes* (New York: Academic Press, 1981), pp. 73–93.

Lorenz, Konrad A., *Man Meets Dog* (London: Penguin, 1953).

Maddock, Janine R., and Shapiro, Lucille, "Polar Location of Chemoreceptor Complex in *Escherichia coli* Cell," *Science* **259** (1993), pp. 1717–1723.

Manney, W. Duntze, and Betz, Richard, "The Isolation, Characterization, and Physiological Effects of the *Saccharomyces cerevisiae* Sex Pheromones," in O'Day , Danton H., and Horgen, Paul A., eds., *Sexual Interactions in Eukaryotic Microbes* (New York: Academic Press, 1981, pp. 21–51).

Meeuse, Bastiaan J.D., "The Voodoo Lily," *Scientific American* **215**(1) (1966), pp. 80–88.

Mitchell, Everett E., "Disruption of Pheromonal Communication Among Coexistent Pest Insects with Multichemical Formulations," *Bioscience* **25** (1975), pp. 493–499.

Miyake, Akio, "Cell Interaction by Gamones in *Blepharisma*," in O'Day, Danton H., and Horgen, Paul A., eds., *Sexual Interactions in Eukaryotic Microbes* (New York: Academic Press, 1981), pp. 95–129.

Moulton, David G., "Communication by Chemical Signals," *Science* **162** (1968), pp. 1176–1180.

Müller-Schwarze, Dietland, Müller-Schwarze, Christine, Singer, Alan G., and Silverstein, Robert M. "Mammalian Pheromone: Identification of Active Component in the Subauricular Scent of the Male Pronghorn," *Science* **183** (1974), pp. 860–862.

Müller-Schwarze, Dietland, and Mozell, Maxwell M., eds., *Chemical Signals in Vertebrates* (New York: Plenum, 1977).

Nichols, James O., *The Gypsy Moth in Pennsylvania*. Miscellaneous Bulletin 4044 (1962).

Parkinson, John S., and Blair, David F., "Does *E. coli* Have a Nose?" *Science* **259** (1993), pp. 1701–1702.

Peters, Roger P., and Mach, L. David, "Scent-marking in Wolves," *American Scientist* **63** (1975), pp. 628–637.

Powers, Bradley, and Winans, Sarah, "Vemeronasal Organ: Critical Role in Mediating Sexual Behavior of the Male Hamster," *Science* **187** (1975), pp. 961–962.

Riddiford, Lynn M. and Williams, Carroll M., "Volatile Principle from Oak Leaves: Role in Sex Life of the Polyphemus Moth," *Science* **155** (1967), pp. 589–590.

Riley, C.V., "Insect Life," *U.S. Division of Entomology* **7** (1994), pp. 33–41.

Rosenblatt, Richard H., and Losey, George S., Jr., "Alarm Reaction of the Top Smelt, *Atherinops affinis*: Reexamination," *Science* **158** (1967), pp. 671–672.

Ross, W.D., *Aristotle Selections* (New York: Scribner, 1955).

Schneider, Dietrich, "The Sex-Attractant Receptor of Moths," *Scientific American* **231** (July 1974), pp. 28–35.

Short, R.V., "Role of Hormones in Sex Cycles," in C.R. Austin and R.V. Short, eds., *Reproduction in Mammals: 3 Hormones in Reproduction* (London: Cambridge University Press, 1972).

Suzuki, Akinori, Mori, Masaaki, Sakagami, Youji, Isogai, Akira, Fujino, Masahiko, Kitada, Chieko, Craig, Ronald A., and Clewell, Don B., "Isolation and Structure of Bacterial Sex Pheromone cPDI," *Science* **226** (1984), pp. 849–850.

Thiessen, Delbert D., "Footholds for Survival," *American Scientist* **61** (1973), pp. 346–351.

Thiessen, D.D., Regnier, Fred E., Rice, Maureen, Goodwin, Michael, Isaacks, Nancy, and Lawson, Nancy, "Identification of a Ventral Scent Marking Pheromone in the Male Mongolian Gerbil (*Meriones unguiculatus*)," *Science* **184** (1974), pp. 83–85.

Todd, John, Atema, Jelle, and Bordach, John C., "Chemical Communication in Social Behavior of Fish, the Yellow Bullhead (*Ictalurus natalis*)," *Science* **158** (1967), pp. 672–673.

Vandenbergh, John G., ed., *Pheromones and Reproduction in Mammals* (New York: Academic Press, 1983).

White, T.H., trans., *The Bestiary, a Book of Beasts* (New York: Putnam, 1960), cited by E.C. Hertzler, "Airborne Pheromones," *Science* **162** (1968), p. 813.

Whitten, W.K., Bronson, F.H., and Grenstein, J.A., "Estrus-inducing Pheromones of Male Mice: Transport by Movement of Air," *Science* **161** (1968), pp. 584–485.

Wilson, Edward O., "Pheromones," *Scientific American* (May, 1963), pp. 100–114.

Wilson, Edward O., *Sociobiology: The New Synthesis* (Cambridge, MA: Harvard University Press, 1975).

Wood, David L., Silverstein, Robert M. and Nakajima, Minoru, *Control of Insect Behavior by Natural Products* (New York: Academic Press, 1970).

Chapter 13

Anon., "A World of Flavors and Scents" Monsanto Magazine, First Quarter (1973).

Bartlett, Des and Jen, "Beavers," *National Geographic* **145** (May, 1974), pp. 716–732.

Bryden, H. A., "The Deer Tribe" in *Mammals of Other Lands* (New York: The University Society, 1917), pp. 245–265.

DeNavarre, Maison G., *The Chemistry and Manufacture of Cosmetics*, 2nd ed., Vol. 1—Background (Princeton: D. Van Nostrand, 1962).

Doty, Richard L., Ford, Mary, Preti, George, and Huggins, George R., "Changes in the Intensity and Pleasantness of Human Vaginal Odors During the Menstrual Cycle," *Science* **190** (1975), pp. 1316–1318.

Gibbons, Boyd, "The Intimate Sense of Smell," *National Geographic* **170** (3), (September, 1986), pp. 324–361.

Gilbert, Avery N., "The Smell Survey Results," *National Geographic* **172** (4), (October, 1987), pp. 514–525.

Lord Haberly, ed., *Pliny's Natural History* (New York: Frederick Ungar Publishing, 1957).

Kelly, Virginia, "The Dollars and Scents of Perfumes," *Reader's Digest* (September, 1974), pp. 54–56.

Kiefer, Otto, *Sexual Life in Ancient Rome* (London: Abbey Library, 1934).

McClintock, Martha K., "Menstrual Synchrony and Suppression," *Nature* **229** (1971), pp. 244–245.

Michael, Richard P., and Keverne, E. B., "Primate Sex Pheromones of Vaginal Origin," *Nature* **223** (1970), pp. 84–85.

Michael, Richard P., Keverene, E. B., and Bonsall, R. W., "Pheromones: Isolation of Male Sex Attractants from a Female Primate," *Science* **172** (1971), pp. 964–966.

Michael, Richard P., Bonsall, R. W., and Warner, Patricia, "Human Vaginal Secretions: Volatile Fatty Acid Content," *Science* **186** (1974), pp. 1217–1219.

Morris, Desmond, *The Naked Ape* (New York: Dell Publishing Co., 1959).

Moulton, David G., "Communication by Chemical Signals," *Science* **172** (1968), pp. 1176–1180.

Poucher, W. A., *Perfumes, Cosmetics and Soaps* (New York: John Wiley & Sons, 1974), revised by A. M. Howard.

Riegel, Emil Raymond, *Industrial Chemistry* (New York: Reinhold, 1942).

C. Plinius Secundus, *The Natural History*, trans. Philemon Holland (New York: McGraw-Hill, 1964).

Semour, George, *Furbearers of California* (Sacramento, CA: California Department of Fish and Game, 1968).

Shreve, R. Norris, *The Chemical Process Industries, 2nd ed.* (New York: McGraw-Hill, 1956).

Stoddard, D. Michael, *The Scented Ape: The Biology and Culture of Human Odour* (Cambridge: Cambridge University Press, 1990).

Thompson, C. J. S., *The Mystery and Lure of Perfume* (London: John Lane The Bodley Head, Ltd., 1927).

Winter, Ruth, *The Smell Book: Scent, Sex, and Society* (New York: J. B. Lippincott, 1976).

Chapter 14

Austin, C. R., and Short, R. V., eds., *Reproduction in Mammals: 3 Hormones in Reproduction* (Cambridge, England: Cambridge University Press, 1972).

Baird, D. T., "Reproductive Hormones," in C. R. Austin and R. V. Short, *Reproduction in Mammals: 3 Hormones in Reproduction* (Cambridge, England: Cambridge University Press, 1972), pp. 1–28.

Berthold, A. A., "Transplantation der Hoden," *Arch. Anat. Physiol. Wiss. Med.* **16** (1849), p. 42., cited by Turner and Bagnara, 1971.

Brown-Séquard, C. E., "Des effets produits chez l'homme par des injections souscutanées d'un liquide retiré des testicules frais de cobaye et de chien." C.r. *Séanc. Soc. Biol.* **I** (1889), pp. 420–430, cited by Goodman and Gilman, 1965.

Butenandt, A., "Über die Chemische Untersuchung der Sexual-hormons," *Z. Angew. Chem.* **44** (1931), p. 44, cited by Goodman and Gilman, 1965.

Calhoun, John B., "Population Density and Social Pathology," *Scientific American* **206** (2) (1962). pp. 139–148, reprinted in *39 Steps to Biology* (San Francisco: W. H. Freeman, 1968), pp. 269–276.

Carlson, G., *The Roguish World of Doctor Brinkley* (New York: Rinehart, 1960).

Christian, John J., "The Effect of Population Size on Adrenal Glands and Reproductive Organs of Male Mice in Populations of Fixed Size," *American Journal of Physiology* **182** (1955), pp. 292–300, reprinted in Richard E. Whalen, *Hormones and Behavior* (Princeton: D. Van Nostrand, 1967), pp. 23–36.

Christian, John J., and Davis, David E., "Endocrine Behavior and Population," *Science* **146** (1964), pp. 1550–1560.

Dimond, Stuart J., *The Social Behavior of Animals* (New York: Harper & Row, 1970).

Erickson, Carl J., and Lehrman, Daniel S., "Effect of Castration of Ring Doves upon Ovarian Activity of Females," *Journal of Comparative and Physiological Psychology* **58** (1964), pp. 164–166, reprinted in Richard E. Whalen, *Hormones and Behavior* (Princeton: D. Van Nostrand, 1967), pp. 49–55.

Feder, Harvey H., and Whalen, Richard E., "Feminine Behavior in Neonatally Castrated and Estrogen Treated Male Rats," *Science* **147** (1965), pp. 306–307, reprinted in Richard E. Whalen, *Hormones and Behavior* (Princeton: D. Van Nostrand, 1967), pp. 182–187.

Ferguson, Marilyn, *The Brain Revolution* (New York: Taplinger, 1973; Bantam Books, 1975).

Fisher, Alan E., "Maternal and Sexual Behavior Induced by Intracranial Chemical Stimulation," *Science* **124** (1956), pp. 228–229.

Freeman, L. V., and Rosvold, H. E., "Sexual Aggressive and Anxious Behavior in the Laboratory Macaque," *Jour. Nerv. Ment. Dis.* **134** (1), (1962), pp. 18–27, cited by Stuart J. Dimond).

Gardner, Martin, *Fads & Fallacies in the Name of Science* (New York: Dover, 1952, 1957).

Getze, George, "A Future Role for Some Old Folk Medicines," quoting Francis Talley of the UCLA Center for Comparative Folklore and Mythology, *Los Angeles Times* (June, 1975).

Gladue, Brian A., Green, Richard, and Hellman, Ronald E., "Neuroendocrine Response to Estrogen and Sexual Orientation," *Science* **225** (September 28, 1984), pp. 1496–1499

Hamer, Dean H., Hy, Stella, Magnuson, Victoria L., Hu, Nan, Pattatucci, Angela, M. L., "Linkage Between DNA Markers on the X Chromosome and Male Sexual Orientation," *Science* **261** (1993), pp. 321–327.

Kolodny, Robert C., Master, William H., Hendryx, Julie, and Toro, Gelson, "Plasma Testosterone and Semen Analysis in Male Homosexuals," *New England Journal of Medicine* **285** (1971), pp. 1170–1174.

LeVay, Simon, *The Sexual Brain* (Cambridge, MA: MIT Press, 1993).

LeVay, Simon, "A Difference in Hypothalamic Structure Between Heterosexual and Homosexual Men," *Science* **253** (1991), pp. 1034–1037.

Lewis, John, *So Your Doctor Recommended Surgery* (New York; Dembuer Books, 1990).

The Merck Index: An Encyclopedia of Chemicals and Drugs (Rahway, NJ: Merck & Co., 1986).

Lillie, F. R., *Jour. Exp. Zool.* 23 (1917), p. 271, cited by R. V. Short, 1972.

Milne, Lorus J. and Margery, *The Biotic World and Man*, 3rd ed. (Englewood Cliffs, NJ: Prentice-Hall, 1965).

Netter, Frank H., *Endocrine System and Selected Metabolic Diseases*, The Ciba Collection of Medical Illustrations, Volume 4 (Summit, NJ: Ciba Pharmaceutical, 1965).

Ryan, Patrick, "Some Prefer Old Age to the Alternative," *Smithsonian* **8** (10), (1978), p. 120.

Schwab, D. F., and Hoffman, M. A., "An Enlarged Suprachiasmatic Nucleus in Homosexual Men," *Brain Research* **537** (1990), p. 141., cited by Marcia Baringa, "Is Homosexuality Biological?" *Science* **257** (1992), pp. 956–957.

Short, R. V., "Sex Determination and Differentiation," in C. R. Austin and R. V. Short, eds., *Reproduction in Mammals: 2 Embryonic and Fetal Development* (Cambridge, England: Cambridge University Press, 1972), pp. 43–71.

Strand, Fleur L., "The Influence of Hormones on the Nervous System," *Bioscience* **25** (9), (1975), pp. 568–577.

Turner, C. Donnell, and Bagnara, Joseph, *General Endocrinology*, 5th ed., (New York: W.B. Saunders, 1971.

Whalen, Richard E., *Hormones and Behavior* (Princeton: D. Van Nostrand, 1967).

Chapter 15

Baird, D. T., "Reproductive Hormones," in C. R. Austin and R. V. Short, eds., *Reproduction in Mammals: 3 Hormones in Reproduction* (Cambridge, England: Cambridge University Press, 1972).

Blount, R. F., "The Effects of Heteroplastic Hypophyseal Grafts upon the Axolotl, *Ambystoma mexicana*," *Jour. Exp. Zool.* **113** (1950), p. 717.

Darwin, Charles, *On the Origin of Species by Natural Selection, or the Preservation of Favored Races in the Struggle for Life* (1859).

Darwin, Charles, *The Descent of Man and Selection in Relation to Sex* (1871), reprinted from the English edition (New York: A. L. Burt Company, 1971).

Dimond, Stuart J., *The Social Behavior of Animals* (New York: Harper & Row, 1970).

Gudernatsch, J. F., "Feeding Experiments in Tadpoles," *Arch. Entwicklungsmech. Organ.* **35** (1912), p. 457, cited by Turner and Bagnara, 1971.

Hayward, S. C., "Modificaton of Sexual Behavior in the Male Albino Rat," *Jour. Comp. Physiol. Psych.* **50** (1957), pp. 70–73, cited by Stuart J.

Dimond, *The Social Behavior of Animals* (New York: Harper & Row, 1970).

Le Boeuf, B. J., "Heterosexual Attraction in Dogs," *Psychonomic Science* **7** (9) (1967), pp. 313–314, cited by Stuart J. Dimond, 1970.

Lehrman, D. S., and Erickson, C., in F. A. Beach, ed., *Sex and Behavior* (New York: John Wiley & Sons, 1962), cited by Stuart Dimond, 1970

Miller, Julie Ann, "The Courtship of Patchwork Flies, *Science* **111** (1977), pp. 107–110.

Paddock, Richard C., and Lynch, Rene, "Surrogate Has No Rights to Child, Court Says," *Los Angeles Times*, May 21, 1993.

Scott, John Paul, *Animal Behavior* (Chicago: University of Chicago Press, 1958, 1971).

Sokolova, L. M., "A Study of Conditional Sexual Reflexes in Rams," *Trud. Lab. Inskusst. Oseme. Zivotn* (Moscow) **1** (1940), pp. 23–35, cited by Stuart J. Dimond, 1970.

Wilson, Glenn, *The Coolidge Effect: An Evolutionary Account of Human Sexuality* (New York: William Morrow, 1982).

Chapter 16

Ardrey, Robert, *African Genesis* (New York: Dell, 1961).

Axelrod, Julius, "The Pineal Gland: A Neurochemical Transducer," *Science* **184** (1974), pp. 1341–1348.

Axelrod, Julius, Velo, G. P., and Fraschini, F., eds., *The Pineal Gland and Its Endocrine Role* (New York: Plenum, 1983).

Ganong, William F., *Review of Medical Physiology* (Los Altos, CA: Lange Medical Publications, 1971).

Fiske, Virginia M., "Serotonin Rhythm in the Pineal Organ: Control by the Sympathetic Nervous System," *Science* **146** (1965), pp. 253–254.

Greiner, A. C., and Chan, S. C., "Melatonin Content of the Human Pineal Gland," *Science* **199** (1978), pp. 83–84.

Hickman, Cleveland P., *Integrated Principles of Zoology* (St. Louis: C. V. Mosby, 1961).

Hoffmann, Roger A., and Reiter, Russel J., "Pineal Gland: Influence on Gonads of Male Hamsters," *Science* **148** (1965), pp. 1609–1610.

Johnson, Carl Hirschie, and Hastings, J. Woodland, "The Elusive Mechanism of the Circadian Clock," *American Scientist* **74** (1986), pp. 29–36.

Klein, David C., and Weller, Joan L., "Indole Metabolism in the Pineal Gland: A Circadian Rhythm in *N*-Acetyltransferase," *Science* **169** (1970), pp. 1093–1095.

Klein, David C., Moore, Robert Y., and Steven M. Reppert, eds., *Suprachiasmatic Nucleus: The Mind's Clock* (New York: Oxford University Press, 1991).

Kuehn, R. E., and Beach, F. A., "Quantitative Measurement of Sexual Receptivity in Female Rats," *Behavior* **21** (3–4) (1963), pp. 282–299, cited by Stuart J. Dimond, *The Social Behavior of Animals* (New York: Harper & Row, 1970).

LeBaron, Ruthann, *Hormones* (New York: Bobbs-Merrill, 1972).

Lewy, Alfred J., Wehr, Thomas A., Goodwin, Frederick K., Newsome, David A., Markey, S. P., "Light Suppresses Melatonin Secretion in Humans," *Science* **210** (1980), pp. 1267–1269.

Luce, Gay Gaer, *Biological Rhythms in Psychiatry and Medicine* (Chevy Chase, MD: National Institute of Mental Health, 1970).

Luce, Gay Gaer, *Body Time: Physiological Rhythms and Social Stress* (New York: Bantam Books, 1971).

Nadler, Ronald D., "Sexual Cyclicity in Captive Lowland Gorillas." *Science* **189** (1975), pp. 813–814.

Reiter, Russel J., *The Pineal Gland* (Boca Raton, FL: CRC Press, 1981).

Scott, John Paul, *Animal Behavior*, 2nd ed., revised (Chicago: University of Chicago Press, 1972).

Simpson, George Gaylord, *The Meaning of Evolution* (New York: Bantam Books, 1971).

Tamarkin, Lawrence, Baird, Curtis J., and Almeida, O. F. X., "Melatonin: A Coordinating Signal for Mammalian Reproduction?' *Science* **227** (1985), pp. 714–720.

Wurtman, Richard J., "The Effects of Light on the Human Body." *Scientific American* **233** (1) (1975), pp.68–77.

Wurtman, Richard J., and Axelrod, Julius, "The Pineal Gland," *Scientific American* **213** (1) (1965), pp. 50–60.

Wurtman, Richard J., Axelrod, Julius, Snyder, S. H., and Chu, Elizabeth W., "Changes in the Enzymatic Synthesis of Melatonin in the Pineal During the Estrous Cycle," *Endocrinology* **76** (1965), pp. 798–800.

Wurtman, Richard J., Axelrod, Julius, and Kelly, D. E., *The Pineal* (New York: Academic Press, 1968).

Young, W. C., ed., *Sex and Internal Secretions* (Baltimore: Williams & Wilkins, 1961).

Chapter 17

Carlson, Elof Axel, *Genes, Radiation and Society: The Life and Work of H. J. Muller* (New York: Cornell University Press, 1981).

Chen, Edwin, "Sperm Bank Donors All Nobel Winners," *Los Angeles Times*, January 29, 1980.

Cohn, Victor, Los Angeles Times-Washington Post News Service, reported in "Sperm-Bank Babies," *Reader's Digest*, (October 1971), pp. 172–173.

Cole, H. H., Cupps, P.T., *Reproduction in Domestic Animals*, 2nd ed. (New York: Academic Press, 1969).

Cupps, P. T., McGown, B., and Rahlmann, D. F., "Semen Quality in Beef Bulls," *California Agriculture* 13 (12), (1959), pp. 5–6.

Edwards, Robert, and Steptoe, Patrick, *A Matter of Life: The Story of a Medical Breakthrough* (New York: William Morrow, 1980).

Goss, Glen W., "Better Mushroom, Hops, Tabasco, and even Mink," in *That We May Eat: Yearbook of Agriculture* (Washington: U.S. Government Printing Office, 1975), pp. 329–336.

Hafez, E. S. E., *Reproduction in Farm Animals*, 2nd ed. (Philadelphia: Lea & Febiger, 1968).

Hardin, Garrett, *Biology Its Principles and Implications* (San Francisco: W. H. Freeman, 1961).

Krier, Beth Ann, "King of the Anonymous Fathers," *Los Angeles Times*, April 21, 1989.

Polge, C., "Increasing Reproductive Potential in Farm Animals," in C. R. Austin and V. Short, eds., *Reproduction in Animals, Book 5: Artificial Control of Reproduction* (London: Cambridge University Press, 1972), pp. 1–31.

Rikmenspoel, Robert, "Biophysical Approaches to the Measurement of Sperm Motility," in David W. Bishop, ed., *Spermatozoan Motility* (Washington: American Association for the Advancement of Science, 1962), pp. 31–54.

Roderick, Kevin, "Rustlers Aim at Vulnerable, Valuable "Seed" of Dairy Herds," *Los Angeles Times*, March 6, 1989.

Chapter 18

Adams, C. E., "Ageing and Reproduction," in C. R. Austin and R. V. Short, *Reproduction in Mammals 4: Reproductive Patterns* (London: Cambridge University Press, 1972), pp. 128–156.

Austin, C. R., "Fertilization," in C. R. Austin and R. V. Short, eds., *Reproduction in Mammals 1: Germ Cells and Fertilization* (London: Cambridge University Press, 1972), pp. 103–133.

Beall, Gary A., "Using Embryo Transfers to Increase Twinning," *California Agriculture* 31 (3), (1977), pp. 8–10.

Berkman, Leslie, "Twins, Grandma? She's Pregnant Pioneer," *Los Angeles Times*, Oct. 6, 1992.

Broad, William J., "A Bank for Nobel Sperm" *Science* 207 (1980), pp. 1326–1327.

David, Georges, and Price, Wendel S., eds., *Human Artificial Insemination and Semen Preservation* (New York: Plenum, 1980).

Edwards, Robert G., "Mammalian Eggs in the Laboratory," *Scientific American* 215 (2), (August 1966), pp. 73–91.

Edwards, Robert, and Steptoe, Patrick, *A Matter of Life: The Story of a Medical Breakthrough* (New York: William Morrow, 1980).

Huxley, Aldous, *Brave New World* (New York: Harper & Row, 1932; Bantam Books, 1953).

Marx, Jean L., "Embryology: Out of the Womb—into the Test Tube," *Science* 182 (1973), pp. 811–814.

New Genetic Technology is Making Investors Bullish on 'Super Cows,' *Los Angeles Times*, November 26, 1984.

Roan, Shari, "A Brave New World?" *Los Angeles Times*, September 10, 1992.

Seidel, George E., Jr., "Superovulation and Embryo Transfer in Cattle," *Science* 211 (1981), pp. 351–358.

Steinbrook, Robert, "No Age Barrier Seen for Egg-Transfer Pregnancies," *Los Angeles Times*, September 9, 1992.

Weil, Jonathan, "CUMC Reports 1st Birth After Embryo Biopsy," *Cornell '93*, (Winter 1993), p. 6.

Winston, Robert M. L., and Handyside, Alan H., "New Challenges in Human in Vitro Fertilization," *Science* **260** (1993), pp. 932–936.

Chapter 19

Abelson, John, "A Revolution in Biology," *Science* **209** (1980), pp. 1319–1321.

Abelson, Philip H., "Improvements in Health Care," *Science* **260** (1993), p. 11.

Amábile-Cuevas, Carlos F., and Chicurel, Marina E., "Horizontal Gene Transfer," *American Scientist* **81** (1993), pp. 332–341.

Baskin, Yvonne, *The Gene Doctors* (New York: William Morrow, 1984).

Baum, Rudy M., "Herbicide-resistant Crops Focus of Biotechnology Debate," *Chemical & Engineering News*, March 8, 1993.

Caplan, A., Herrera-Estrella, L., Inzé, D., Van Haute, E., Van Montagu, M., Schell, J., Zambryski, P., "Introduction of Genetic Material into Plant Cells," *Science* **222** (1983), pp. 815–821.

Carlson, Elof Axel, *Genes, Radiation and Society: The Life and Work of H. J. Muller* (Ithaca: Cornell University Press, 1981).

Cohen, Stanley N., "The Manipulation of Genes," Scientific American **233** (1), (July, 1975), pp. 24–33.

Davis, Bernard D., ed., *The Genetic Revolution: Scientific Prospects and Public Perceptions* (Baltimore: Johns Hopkins University Press, 1991).

Elkington, John, *The Gene Factory: Inside the Genetic and Biotechnology Business Revolution* (New York: Carroll & Graf, 1985).

Gordon, Jon W., and Ruddle, Frank H., "Integration and Stable Germ Line Transmission of Genes into Mouse Pronuclei," *Science* **214** (1981), pp. 1244–1246.

Graf, Lloyd H., Jr., "Gene Transformation," *American Scientist* **70** (1982), pp. 496–505.

Gribbin, John F., *In Search of the Double Helix: Quantum Physics and Life* (New York: McGraw-Hill, 1985).

Grobstein, Clifford, "Recombinant-DNA Debate," *Scientific American* **237** (1), (1977), pp. 22–33.

Hall, Stephen S., *Invisible Frontiers: The Race to Synthesize a Human Gene* (New York: Atlantic Monthly Press, 1987).

Hileman, Bette, "FDA Panel Okays Bovine Growth Hormone," *Chemical & Engineering News*, April 5, 1993.

Jacobs, William R., Jr., Barletta, Raúl G., Vdami, Rupa, Chan, John,

Kalkut, Gary, Sosne, Gabriel, Kieser, Tobias, Sarkis, Gary J., Halfull, Graham F., Bloom, Barry R., "Rapid Assessment of Drug Susceptibility of *Mycobacterium tuberculosis* by Means of Luciferase Reporter Phages," *Science* **260** (1993), pp. 819–822.

Kenny, Martin, *Biotechnology: The University-Industrial Complex* (New Haven: Yale University Press, 1986).

Lee, Thomas F., *The Human Genome Project: Cracking the Genetic Code of Life* (New York: Plenum, 1991).

Marx, Jean L., "Building Bigger Mice Through Gene Transfer," *Science* **218** (1982), p. 1298.

Mayr, Ernst, *The Growth of Biological Thought: Diversity, Evolution and Inheritance* (Cambridge: Harvard University Press, 1982).

Palmiter, Richard D., et al., Metallothionein-Human GH Fusion Genes Stimulate Growth of Mice," *Science* **222** (1983), pp. 809–814.

Reisch, Marc, "Insect-killing Cotton Passes Field Test," *Chemical & Engineering News*, Oct. 29, 1990.

Sayre, Anne, *Rosalind Franklin and DNA* (New York: W. W. Norton, 1975).

Smollar, David, "Tobacco Lights Up, Providing a Method for Tracing Genes," *Los Angeles Times*, November 7, 1986, p. 3.

Stolberg, Sheryl, "New, More Expensive Heart Drug Outpaces Old Medicine in Test," *Los Angeles Times*, May 1, 1993.

Sylvester, Edward J., and Klotz, Lynn C., *The Gene Age* (New York: Scribner, 1983).

Thayer, Ann M., "Poor 1992 Earnings, Shaky Stocks Plague Biotechnology Firms," *Chemical & Engineering News*, March 29, 1993, pp. 17–19.

Watson, James D., *The Double Helix* (New York: Van Nostrand Reinhold, 1983).

Watson, James D., and Crick, Francis H. C., "Molecular Structure of Nucleic Acids: A Structure for Deoxyribonucleic Acid," *Nature* **171** (1953), pp. 737–738.

Watson, James D., Gilman, Michael, Witkowski, Jan, and Zoller, Mark, *Recombinant DNA*, 2nd ed. (New York: W. H. Freeman, 1992).

Weaver, Robert F., "Changing Life's Genetic Blueprint," *National Geographic* **166** (6), (Dec. 1984), pp. 818–847.

Wetzel, Ronald, "Applications of Recombinant DNA Technology," *American Scientist* **68** (1980), pp. 664–675.

Zimmerman, Burke, *Biofuture* (New York: Plenum, 1984).

Index

311